CAMBRIDGE LIBRARY COLLECTION

Books of enduring scholarly value

Monographs of the Palaeontographical Society

The Palaeontographical Society was established in 1847, and is the oldest Society devoted to study of palaeontology worldwide. Its primary role is to promote the description and illustration of the British fossil flora and fauna, via publication of an authoritative monograph series. These monographs cover a wide range of taxonomic groups, from microfossils, trilobites and ammonites through to Coal Measure plants, mammals and reptiles, and from all ages from Cambrian to Pleistocene. They form a benchmark for understanding the past life of the British Isles and many include the original descriptions of numerous key species. The first monograph (on the Crag Mollusca) was published in March 1848 and the Society still continues this work today. Notable authors in the series include Charles Darwin (fossil barnacles) and Richard Owen (dinosaurs and other extinct reptiles). Beginning in 2014, the Cambridge Library Collection and the Society are collaborating to reissue the earlier publications, focusing on monographs completed between 1848 and 1918.

A Monograph of the Crag Mollusca

The Pliocene–Pleistocene Crags of East Anglia are an incredibly rich source of fossil shells, many belonging to extant Boreal and Mediterranean genera. Dominated by marine gastropods and bivalves, the deposits also contain evidence of terrestrial and non-marine gastropods and bivalves, brachiopods, and extensive epifauna including bryozoans. Published between 1848 and 1879 in four volumes, the latter two being supplements with further descriptions and geological notes, this monograph by Searles Valentine Wood (1798–1880) covers more than 650 species and varieties of fossil mollusc. For each species Wood gives a synonymy, diagnosis (in Latin), full description, dimensions, occurrence and remarks. The supplements also provide a breakdown of the species and their current distribution. The detailed plates were prepared by the conchologist George Brettingham Sowerby and his namesake son. Volume 4 (1879) comprises the second supplement, covering univalves and bivalves. Also included here is a third supplement, published by the late author's son in 1882.

Cambridge University Press has long been a pioneer in the reissuing of out-of-print titles from its own backlist, producing digital reprints of books that are still sought after by scholars and students but could not be reprinted economically using traditional technology. The Cambridge Library Collection extends this activity to a wider range of books which are still of importance to researchers and professionals, either for the source material they contain, or as landmarks in the history of their academic discipline.

Drawing from the world-renowned collections in the Cambridge University Library and other partner libraries, and guided by the advice of experts in each subject area, Cambridge University Press is using state-of-the-art scanning machines in its own Printing House to capture the content of each book selected for inclusion. The files are processed to give a consistently clear, crisp image, and the books finished to the high quality standard for which the Press is recognised around the world. The latest print-on-demand technology ensures that the books will remain available indefinitely, and that orders for single or multiple copies can quickly be supplied.

The Cambridge Library Collection brings back to life books of enduring scholarly value (including out-of-copyright works originally issued by other publishers) across a wide range of disciplines in the humanities and social sciences and in science and technology.

A Monograph of the Crag Mollusca

With Descriptions of Shells from the Upper Tertiaries of the East of England

VOLUME 4: SECOND AND THIRD SUPPLEMENTS
(UNIVALVES AND BIVALVES)

SEARLES V. WOOD

ORDER OF BINDING AND DATE OF PUBLICATION OF VOLUME IV.

PAGES	PLATES	ISSUED IN VOL. FOR YEAR	PUBLISHED
Title-page; Preface; 1—58	I—VI	1879 „ „	May, 1879 „ „

SECOND SUPPLEMENT

TO THE

MONOGRAPH OF THE CRAG MOLLUSCA,

WITH

DESCRIPTIONS OF SHELLS

FROM THE

UPPER TERTIARIES OF THE EAST OF ENGLAND.

BY

SEARLES V. WOOD, F.G.S.

VOL. IV.

UNIVALVES AND BIVALVES.

LONDON:
PRINTED FOR THE PALÆONTOGRAPHICAL SOCIETY.
1879.

PRINTED BY
J. E. ADLARD, BARTHOLOMEW CLOSE.

PREFACE.

When I had completed my first Supplement to the " Crag Mollusca " in 1872–4, I did not contemplate ever attempting any further addition, as even if I had desired to make any, my advanced years rendered it improbable that I could accomplish such a thing. The discovery, however, of some shells at Boyton, one of them (*Fusus Waelii*) apparently identical with a shell from older beds in Belgium and Germany, and two others (*Murex Reedii*, and *M. pseudo-Nystii*) presenting an approach to certain Murices of the same older beds, were of such interest as to render their representation by figure and description desirable, for if, as is probable, they lived in the Coralline Crag sea, they furnish evidence of a nearer connection of that sea with the Miocene than modern opinion has been inclined to grant.

I was thus induced to enter upon a second Supplement, which I at first thought might be confined to a single plate, but when this had been engraved I reflected that as so many species had been introduced into lists of Crag shells, which I had not introduced into my first Supplement from a feeling that the authority for them was too scant or doubtful to justify it, or, in some instances, from a feeling that the identity was erroneous, it was incumbent on me to present to geologists by figured representations the evidence upon which these introductions were based. This, therefore, I have endeavoured to do, and by it have, perhaps, exposed myself to the objection that the plates have been extended to but little purpose, as many of the so-called new species are either very doubtful in themselves, or are merely derivatives from destroyed beds; though most of these beds probably belong either to the Coralline, or to some still older part of the Crag; *i.e.* to the oldest Pliocene, now present in Belgium. To such objections my answer would be that I have long felt that the introduction of so many new species into Crag lists, either from the unsatisfactory evidence of a single specimen, or from the (in my view) improper identification made, or from the presence of mere derivatives, must produce among geologists, especially those abroad, very erroneous conceptions of the Crag Fauna; and that it was to the advantage of science that these evidences should be placed in an appreciable form before the scientific world.

I fear that most of the additions thus made of late years to the Crag Fauna, coupled with the antagonism between the views of Dr. Jeffreys, concerning the identification of many Crag shells with recent species (as expressed by the list which accompanies the paper of Prof. Prestwich, in the twenty-seventh Volume of the 'Journal of Geological Society') and those of myself, will render the subject of the Crag Mollusca, for some time to come, a subject of more perplexity than interest to students of the upper tertiaries.

I have now by inquiry in every quarter which afforded the slightest chance of result exhausted all possible additions to the Molluscan Fauna of the Crag up to the present time, doubtful or otherwise, and dealt with them in the present Supplement.

Dr. Lycett has (after a lapse of more than twenty years) written to me that the attribution of an analysis of the *Myadœ* to Prof. Morris made in the footnote to p. 265 of my second volume of the " Crag Mollusca " was an error, and that the analysis was entirely his own. I take this opportunity, therefore, of acknowledging the error, and of expressing my regret for it.

S. V. WOOD.

November, 1873.

SECOND SUPPLEMENT

TO THE

CRAG MOLLUSCA.

BUCCINUM NUDUM, *S. Wood.* 2nd Sup., Tab. I, fig. 1 *a, b.*

Spec. Char. B. *Testá tenui, elongato-ovatá, turritá, lævigatá, apice obtusá, depressá; anfractibus septenis, convexiusculis; suturá impressá; aperturá ovatá; labro tenui acuto, columellá regulariter concavá.*

Axis 2¼ inches.

Locality. Cor. Crag, Sutton.

The shell here represented is from the collection of Mr. Canham, who tells me he obtained it from the lower part of the Cor. Crag at Sutton. The shell is very thin and fragile and has lost some small portion of its exterior and a small part of the shell, but it has retained its natural form by the somewhat slight consolidation of the material within. It resembles a shell I figured in my Suppl., Addendum Plate, fig. 11, under the name of *Buc. Tomlinei,* but that is not quite so elongated as the present one, and it is ornamented with large and distinct spiral striæ; while our present shell, where the outer coat has been preserved, appears to have been perfectly smooth and very thin. I have a cast of this shell in one of the so-called "box stones" of the Red Crag. It belongs apparently to a group of shells of which *Buc. Dalei* may be considered as the type; but it departs as much or more from that species as does the other Cor. Crag shell *pseudo-Dalei.* Both, however, are obnoxious to the same objection that they are founded on solitary specimens. To this objection the extreme rarity in the Cor. Crag of the normal form *Dalei* is to some extent an answer.

At fig. 5 *a, b,* tab. i, of the same plate is represented a specimen which I have referred (with doubt) as a deformity to *Buc. undatum;* it somewhat resembles a shell I figured in Sup. to Crag Moll., tab. ii, fig. 5, and considered as a deformed specimen or variety of that species, and I am inclined to think our present shell is in a similar condition. It was sent to me by Dr. Reed, and is said to have come from the Red Crag of Butley, the locality from which I obtained my specimen. The volutions are somewhat angulated at

1

the base, and slightly so at the shoulder, where there are traces of undulated ridges like those of *undatum*.

I have also figured another shell from the Cor. Crag belonging to Dr. Reed which, I think, is a deformed specimen of *Buccinum Dalei* (2nd Sup., tab. i, fig. 2); the thickened margin was formed, I imagine, when its growth was arrested, and the ridge upon the columella is, I think, the result of disease, and therefore only a malformation.

BUCCINUM DECLIVE, *S. Wood.*   2nd Sup., Tab. II, fig. 10 *a, b.*

Locality.   Cor. Crag?   Boyton.

This is another specimen out of the rich cabinet of Dr. Reed, who gives it from that somewhat doubtful locality of Boyton.   This specimen may be described as ovato-fusiformi, spirâ elevatâ, apice obtusâ, spiraliter striatâ, anfractibus 5—6 convexis, suturis depressis, valdè distinctis, obsoleté costatâ; aperturâ ovatâ, labro simplici acutâ; canali breve.   It is, I believe, distinct from any of the varieties of the variable shell *B. undatum*, the volutions are more convex, with a much deeper suture, and it has a more obtuse or mammillated apex.

The shell has been a good deal rubbed.   The striæ, although somewhat obliterated, are visible in places, and the longitudinal ridges are also visible, but not very regular or distinct.   These do not appear to be at all " undulated " as if the outer lip had been sinuated, and as this character seems to indicate that the shell is distinct from *undatum*, I have assigned to it the above name, but it must be regarded as a doubtful species.

NASSA PRISMATICA, *Brocchi.*   2nd Sup., Tab. I, fig. 6.

BUCCINUM PRYSMATICUM, *Broc.*   Conch. Foss. Subap., p. 337, t. v, fig. 7, 1814.

Spec. Char.   " *Testá ovato-oblongá, longitudinaliter costatá, striis transversis crebris, elevatis, labro columellari, supernè uniplicato, basi reflexá, emarginatá* " (Brocchi).

Axis 1 inch.

Localities.   Cor. Crag, Sutton.

Fossil in Piacentino, Italy.

The present specimen is from the cabinet of the Rev. Mr. Canham, and from the lower part of the Coralline Crag.   The shell represented under this name in the Crag Moll. vol. i, p. 32, tab. iii, fig. 6, is, I now believe, a distinct species, and I have resumed the name of *Nassa microstoma* for it as next described.

Our present specimen is not quite so large as the one figured by Brocchi, which is a full-grown shell, whereas the one now represented has not attained to maturity, and has the outer lip sharp without denticulation on the inside of it.

NASSA MICROSTOMA, *S. Wood.* 2nd Sup., Tab. I, fig. 4 *a, b.*

> NASSA MICROSTOMA, *S. Wood.* Catal. Mag. Nat. Hist., 1842.
> — PRISMATICA, *S. Wood.* Crag Moll., vol. i, p. 32, t. iii, fig. 6, 1848.
> — ELEGANS, *Dujard.* Tr. Geol. Soc. Fr., p. 298, pl. xx, figs. 3—10, 1837.

Spec. Char. *Testá turritá, spirá elevatá, costatá, costis* 20—24, *spiraliter striatá ; anfractibus* 7—8, *convexis, suturis profundis, aperturá rotundato-ovatá ; labro incrassato, intùs denticulato ; labio supernè uniplicato.*

Axis $\frac{9}{16}$ of an inch.

Locality. Cor. Crag? Boyton.

Fossil in Touraine, France.

The specimen represented in the above figure is from the cabinet of Mr. Robert Bell, and he tells me that it came from Boyton. Doubts occur as to whether shells from this locality, not previously known in the Crag, belong to the Red or to the Coralline Crag,[1] but I am inclined to refer our present specimen to the older formation, both from the colour and appearance of the shell and from its apparent connections.

I now consider this species as specifically distinct from *prismatica,* and probably the same as the shell figured in Crag Mol., vol. i, Pl. III, fig. 6, and which in my synoptical list is inserted as *Nassa prismatica* var. *limata.* I refer it to *N. elegans,* Dujardin, an abundant Touraine shell which is much less than *prismatica,* has a greater number of costæ, and a smaller opening comparatively ; as it is quite distinct from the well-established Red Crag species called *N. elegans* by the late Rev. G. R. Leathes in 1824, while Dujardin's name of *elegans* bears a date of 1837, it is necessary to suppress the latter to avoid confusion, and I have therefore assigned to it the name *microstoma* which I used first in my catalogue of 1842 referred to.

[1] I have not been able to see the Boyton excavation open, but I am informed that a thin layer of Red Crag is found there reposing upon a small thickness of Coralline, and the whole being inundated with water the two are shovelled out together and washed for the phosphatic nodules, so that the specimens from each bed are intermingled beyond possibility of distinction other than what may be drawn from the appearance of the specimen or the character of the species.

NASSA CONSOCIATA, *S. Wood.* 2nd Sup., Tab. IV, fig. 13 *a, b*; Crag Moll., vol. i, p. 31, Tab. III, fig. 7.

Axis ¾ths of an inch.

Locality. Red Crag, Waldringfield.

The specimen figured as above referred to is said by Mr. Canham to be from Waldringfield, and is in the collection made by him and now placed in the Ipswich Museum. That locality has yielded so many derivatives that I think the present shell may have been introduced from the destruction of material belonging to the Coralline Crag period. It is larger than any specimen I have from this latter formation, but this constitutes the only difference that I can discover.

Tab. IV, fig. 15, represents a small specimen of *Nassa* from the Red Crag of Butley, sent to me by Mr. Robert Bell with the MS. name of *N. tumida,* as he considers it a distinct species. This I have had figured, as it presents some differences from *N. incrassata* (the shell to which I believe it approaches nearest) in being more ovate and possessing more numerous costæ, and in being smaller; but as I do not think that these suffice to distinguish the shell specifically from *incrassata,* I have here called it var. *tumida* of that species. In the same Plate, fig. 12, is represented a small specimen from the Red Crag of Sutton, which I think is only a dwarf individual of *Nassa granulata,* here called var. *nana;* it much resembles *N. granifera,* but in that shell the costæ stand further apart with a plain space between them. In our present shell the costæ meet at the bases.

NASSA ANGULATA ? *Brocchi.* 2nd Sup., Tab. IV, fig.

BUCCINUM ANGULATUM, *Broc.* Conch. Foss. Subap., p. 654, tab. xv, fig. 18, 1814.

Locality. Boyton.

This is another form of the genus *Nassa* for which I have had great difficulty in making a reference, and have given to it the above one provisionally, having seen but the single specimen now figured, and this comes from a locality of doubtful age. It is from Mr. Robert Bell.

COLUMBELLA ? (ASTYRIS) SULCULATA, *S. Wood.* 2nd Sup., Tab. I, fig. 3.

Spec. Char. C. Testá turritá, elongatá, spirá elevatá, apice obtusá, acuto? anfractibus convexiusculis, transversim late sulcatis; aperturá quadrato-ovatá; labro intùs denticulato; basi truncatá, canali breve.

Axis $\frac{3}{4}$ths of an inch.

Locality. Red Crag, Sutton, Shottisham.

The specimen figured is from the cabinet of Dr. Reed, and to this the name of *Lachesis magna* was attached by Mr. A. Bell, but it appears to me to approach so near to *Columbella sulcata*, J. Sow., from Walton Naze, see Crag Moll., vol. i, p. 23, tab. ii, fig. 2, that I have given to it the same generic name of that aberrant section of *Columbella*.

Our present shell may be described as having an elevated spire, volutions slightly convex, ornamented with five or six rather broad and flattened striæ, separated by a fine and narrow line, with a deep and distinct suture; the aperture is ovately quadrangular, but not so much so as that of *C. sulcata*; the columella somewhat concave, and the canal short; the apex is not quite perfect.

Since the figure was engraved Mr. Robert Bell has presented me with a specimen of this species, a trifle larger than the one figured, and to this he has given the generic name of *Pisania*, but I see nothing in the specimen to require (according to my view) a new generic position.

I have here also given the representation of a shell in my own cabinet (2nd Sup., tab. iii, fig. 11), which I think is a distorted, abraded, and immature specimen of *Columbella sulcata*. It is ornamented with the same kind of spiral striæ, the last whorl (only) inflated, and the volutions are made more convex by decortication.

Lachesis Anglica, Sup., Crag Moll., Addendum Plate, fig. 7, probably belongs to the same section of *Columbella*. I do not know what especial character is given to the shell for the generic name of *Lachesis*.

PURPURA LAPILLUS. 2nd Sup., Tab. I, fig. 13.

The shell shown in the above figure represents a specimen that has been sent to me with the name of "Buccinum?" but I believe it to be simply a distortion of *Purpura lapillus*, and as it comes from Bramerton, whence I had previously received many specimens of other shells greatly distorted, I am strengthened in this view, and the shell may be classed with other distorted specimens figured in the Crag Moll.; see tab. iv, fig. 6, and tab. xix, fig. 12. The full-grown individuals of this species, or at least nearly all of them, have the outer lip sharp and simple, but in the young state the specimens are sometimes regularly and strongly dentated on the inside of the outer lip. I have other specimens of the same size, and less than the one figured, which have a few and strongly marked denticles on the right side of the aperture, but in general they are absent. The present specimen has been much rubbed and abraded, indicating the shallowness of the water in which it had lived. What should cause this peculiar dentation to the aperture in some of the young shells and not in others I am unable to explain. This character of dentation is an accompaniment of the full-grown shell in most species rather than of the young, and I have had the specimen figured lest by any chance it should have been regarded as some new species and added to the number of such in lists of crag shells for which I can find no warrant.

Captain Brown has figured a specimen of this species with a dentated outer lip ('Illustr. Conch. Grt. Britain,' Pl. xlix, fig. 6), which he has called *Purpura Anglicanæ*, referring to 'Lister's Conch.,' Pl. 965, fig. 18. "Lister does not say from whence he obtained this singular variety" (Brown).

TROPHON (SIPHO) ISLANDICUS, *Chemnitz*.   2nd Sup., Tab. II, figs. 3 *a*, 3 *b* recent.

FUSUS ISLANDICUS, *Forb. and Hanl.*   Brit. Moll., vol. iii, p. 416, pl. ciii, fig. 3, 1853.

Locality.   Red Crag, Sutton.

The shell figured as above represents a specimen which I found many years ago and regarded as a var. of *Trophon gracilis*, figured and described in Crag Moll., vol. i, p. 46, tab. vi; but which I here give as a true representation of the recent British shell called *Islandicus* (fig. 3 *a*) ; and by the side of it have had engraved the figure of a recent specimen of that species for comparison,(fig. 3 *b*) because it has been said not to be a crag species.   This shell is rather more elongated than *gracilis*, and deserves the name of *angustius*, originally given to it long before the time of Linné or of Gmelin, and which I adopted in my original catalogue published in the Annals of Nat. Hist. in 1842, p. 541. That name, however, being anterior to the time of our starting point, the 12th edit. of Linné, I give the shell under the usually received name of *Islandicus*.

TROPHON (SIPHO) TORTUOSUS, *L. Reeve*.   2nd Sup., Tab. II, fig. 2 *a, b*.

TROPHON GRACILE, var. *S. Wood*.   Crag Moll., vol. i, p. 46, tab. vi, fig. 10 *b*, 1848.

Dr. Reed has lately sent me several specimens both from the Coralline and Red Crags that belong to a group of shells of which *Fusus Islandicus* may be considered as the type. Among those from the Red is one (fig. 2*a*) supplied by Mr. A. Bell and marked by the latter as *Fusus tortuosus* of L. Reeve, figured and described in Sir Edward Belcher's 'Last of the Arctic Voyages,' vol. ii, p. 394, Pl. xxxii, fig. 5 *a, b*.

The shell figured in the Crag Moll., tab. vi, fig. 10 *b*, is referred by Mr. A. Bell to the same species, and I am now disposed to think that Mr. Bell's references of this shell to Lovell Reeve's species is correct, if the differences be sufficient to constitute a specific removal. Mr. Bell also says that fig. 10 *a*, *c*, of the same plate may be referred to *Fusus Olavii*, Beck, and considered a distinct species.

The principal character, indeed I believe the only one, by which *tortuosus* can be distinguished from either *gracilis* or *propinquus* is the greater convexity of the volutions; the form of the canal being similar in each with the volutions covered by regularly broad-spiral striæ. I have here had represented as above (fig. 2 *a*) the specimen from Dr. Reed, and which, in outward form, varies from the figure in the Crag Moll. as also from that given as mentioned by Lovell Reeve. I think it may be considered only as a variety; it is said to have come from Waldringfield. Fig. 2 *b* of my present plate is the representation of a specimen of my own found by myself in the Red Crag at Sutton many years ago, and this I now think is only a slight distorted form of *tortuosus*, as I have two others similar in the volutions, but not so perfect, and thought it only a variety, not of sufficient importance to deserve a figure; but so many separations having been made out of a group of shells which probably may be united under the name of *Sipho*, I have had it here figured and have endeavoured to group these shells together under that name, which have been found in the Upper Tertiaries of the east of England, viz. :

Trophon (Sipho) Islandicus? *Chem.* 2nd Sup., tab. ii, fig. 3. Red Crag.
—   —   Olavii, *Beck.* Crag Moll., vol. i, p. 46, tab. vi, fig. 10 *a, c.* Red Crag.
—   —   gracilis, *Da Costa.* 2nd Sup., tab. ii, fig. 4. Cor. Crag.
—   —   propinquus, *Alder.* App. Crag Moll., tab. xxxi, fig. 3 *a. b.* Cor. Crag.
—   —   id.   Sup., tab. vii, fig. 21, sinistral. Red Crag.
—   —   id.   2nd Sup., tab. ii, fig. 5. Cor. Crag.
—   —   Sarsii, *Jeff.* Sup., p. 23, tab. i, fig. 9. Red Crag.
—   —   tortuosus, *L. Reeve.* Crag Moll., vol. i, tab. vi, fig. 10 *b.* Red Crag.
—   —   id.   Sup., tab. ii, fig. 15 *a.* Red Crag.
—   —   id.   2nd Sup., tab. ii, fig. 2 *a, b.* Red Crag.
—   —   Sabini, *Hancock.* Sup., tab. ii, fig. 15 *c.* Bridlington.
—   —   ventricosus, *Gray.* Sup., p. 22, tab. iii, fig 4. Bridlington
—   —   Leckenbyi, *S. Wood.* Sup., p. 24, tab. vii, fig. 1. Bridlington.

The whole of these may very probably be only inconstant varieties of *Islandicus*, but I have figured them under the names of their authors to show their occurrence in the deposits embraced by my Monograph. *T. Leckenbyi* of myself stands in this respect on an equal footing with the other so-called species given above.

Note.—*Sipho*, Klein, 1753. This name is previous to our starting point, the 12th edit. of Linné, but it appears now to be adopted by many of our conchologists.

TROPHON PSEUDO-TURTONI, *S. Wood.* 2nd Sup., Tab. II, fig. 1 ; and Tab. IV, fig. 1.

> TROPHON NORVEGICUS ? *Chemn.* Appendix to Crag Moll., t. xxxi, fig. 1 ; 1st Supplement to Crag Moll., t. v, fig. 14 ; and Addendum Tab., fig. 16.

Locality.—Red Crag, Waldringfield.

In the Appendix to the Crag Mollusca and in my previous Suppl. are figured and described some specimens of this shell, none of them perfect, under the name of *Trophon Norvegicus.* The perfect specimens which I am now able to represent seem to me to differ so considerably, however, from the recent shell called *Norvegicus,* that I have proposed for it the above name, indicative at once of its distinctive character from *Norvegicus* and of its affinity to that species. Our present shell possesses more convex volutions and a much deeper suture, a longer spire with a smaller and shorter opening. The recent shell *Norvegicus* is described as having " the body whorl disproportionately large compared with the spire ;" " the body occupies ⅗ths of the dorsal length." The body whorl of our present fossil measures only half of its entire length, and is also more strongly striated ; for assuming even that it has been decorticated and lost some of its outer coating, these striæ are more visible than those on the living shell, which on a specimen in my possession are principally confined to the epidermis, or at least are but very slightly visible beneath it. I am anxious to have this fossil correctly described and delineated because in a list of fossils from Uddevalla, by Mr. Jeffreys, read at the Brit. Assoc. 1863, at p. 77, is the name of *Fusus Turtonii,* Bean, with this remark "a var. approaching in shape *F. Norvegicus;*" and I imagine this Uddevalla fossil may possibly be the same as our present specimen. I cannot, however, fairly refer the shell figured to either of those species ; and it appears to me to be intermediate between the two. The late Dr. S. P. Woodward in his list of shells from the Norwich crag has the name of *T. Norvegicus* (J. M. and R. F.) which as well as the one called by Mr. Bell *F. Lagillierti* (Sup. to Crag Moll., Addendum Plate, fig. 16), may also, I imagine, be the same as the present shell.

The specimen figured, Tab. IV, fig. 1, is from the Ipswich Museum by the kindness of Dr. J. E. Taylor, the curator.

TROPHON (TRITONOFUSUS) ALTUS, *S. Wood.* 2nd Sup., Tab. I, fig. 11. Crag Moll., vol. i, Tab. VI, fig. 13, as *Trophon altum.* 1st Sup., p. 23, Tab. II, fig. 17.

To whatever genus this shell may belong, the specimens exhibit great variation like those of *Buccinum undatum* and *Trophon antiquus*. A further figure which I have now given shows the canal not to be prolonged beyond the lower portion of the outer lip, corresponding in that respect to the diagnosis of the genus *Buccinum*. Some of the specimens I have figured and referred to this species have on the upper portion of the spire some obsolete costæ, which are absent from our present specimen; but this, I think, is insufficient for specific removal, as the same differences may be seen in specimens of the common *Buc. undatum*.

The specimen now figured is from the cabinet of Dr. Reed, who obtained it from Mr. A. Bell, by whom it had been labelled as a new species from the Red Crag, Butley, which was one of the reasons that induced me to have it figured. It is a very perfect specimen, and shows an expanded lip like that of *Buccinum*.

TROPHON (BUCCINOFUSUS) KRÖYERI ? *juv. Moller.* 2nd Sup., Tab. III, fig. 8.

FUSUS KRÖYERI, *Möll.* Index Moll. Grœnlandiæ, p. 15, 1842.

Axis 1 inch.

Locality. Red Crag, Shottisham.

The present specimen has been sent to me by Mr. Robert Bell with the above name, and I give it on his authority; he says he has compared it with a recent specimen of the above name in the British Museum, and it appears to him to correspond with the younger or upper part of that species. I saw that species in the British Museum some years ago, and so far as my memory will assist me, I think probably it may be so. I have given to it the above name with a mark of doubt, as it will be necessary to have a better specimen for a more correct determination. The specimen is without striation, or otherwise the striæ have been obliterated.

FUSUS WAELII, *Nyst.* 2nd Sup., Tab. I, fig. 10 *a, b, c.*

FUSUS WAELII, *von Könen.* Mitt. Oligoc., p. 76, taf. vi, fig. 2 *a—d*, 1867.
— — *S. Wood.* Quart. Journ. Geol. Soc., vol. xxxiii, p. 120, 1877.

Spec. Char. F. Testá elongato-fusiformi, spirá elevatá, apice obtusá; anfractibus convexis, longitudinaliter costatis, spiraliter striatis; aperturá ovatá; canali, elongato paulo contorto terminato.

Axis, 1 inch.

Locality. Cor. Crag, ? Boyton.

This shell was noticed by me as from the Coralline Crag in the 'Quart. Jour.' of the Geol. Soc. above referred to, and I have now the opportunity of figuring the specimens. I have also since then received two specimens of the typical oligocene form from Dr. Nyst, from the locality of Baesele, near Boom (Rupelien); and I think the British Crag Fossil may safely be referred to it. The only difference which I can detect is that the inside of the outer lip in one of the Belgian specimens is denticulated, while that of the Crag shell is not. The other specimen sent to me by Dr. Nyst, however, does not present this character; nor so far as I can see do specimens sent me by Dr. Von Könen, from the German Oligocene of Sternberger Gastein, nor by some specimens from the Oligocene of Rupelmonde, in Belgium, sent me by M. Rutot; the artist has given a representation to my specimen which might be mistaken for denticulations on the inside of the outer lip, but there are none, and the ribs are not so wide and coarse as he has shown them. I have had the only two specimens (which I believe have as yet been found) figured, one of which is more elongated than the other, and they appear to correspond as well with the two figures given by Dr. Von Könen as with the oligocene specimens to which I have referred. Our shell has eight, somewhat rounded ribs or costæ upon the last volution, the spiral striæ resemble those upon the Baesele shell, and the caudal termination is long and slightly twisted as in the one before mentioned; the apex is obtuse, with the first volution apparently smooth, but the volution not being perfect this cannot positively be affirmed. This shell also very strongly resembles *Fusus crispus*, and a worn specimen was figured by me in my first Suppl. under that name, with a note doubting the correctness of the reference (p. 29, Tab. II, fig. 10). Two specimens with the name of *F. crispus*, Broc., and the syn. *F. Rothi*, and the locality Bekken (miocene) attached, I have, by the kindness of M. Bosquet, long possessed, and these show prominent and sharp spiral striæ, with two small ridges upon the columella; but these ridges are not visible in the only two worn specimens from the Crag, on which I made the reference in p. 29 of my Suppl. A fine specimen of *F. crispus*, Borson, sent me by Dr. Von Könen, from the Miocene of Langenfelde near Hamburg, has the inner part of the outer lip denticulated, but has no folds on the columella; in other respects it agrees with specimens sent me from the bed at Kiel and Edeghem in Belgium, under the name *F. sexcostatus*. A specimen of *F. sexcostatus* from the Miocene of Dingden near Wesel, kindly sent me by Dr. Könen is destitute of these folds on the columella, and were it not that the three upper whorls are smooth (which is not the case with the Crag specimens), would equally agree with the more elongated form of the two now given specimens figured above. On the other hand, specimens sent me by M. Rutot, under the name of *F. sexcostatus*, from the so-called Miocene of Kiel and Edeghem in Belgium, with the apices perfect, are destitute of these three unornamented whorls; but one of them has two folds on the columella; another (the largest) has but one, while another, the smallest, has none at all. Not one of these three last-mentioned specimens has the inside of the outer lip denticulated, and the smallest of them is not distinguishable in any respect from the longer of the two Crag specimens which I have figured under the name of *Waelii*. Under these circumstances it seems to me that, though *F. Waelii* is not recognised as a species of the Belgian Miocene (a formation which M. Vanden Broeck now refers to the oldest Pliocene, contending that the true Miocene is absent in Belgium), the shell I have figured under this name does occur in the Belgian formations; and it may perhaps be that, if a large series of specimens of *F. Waelii*, *F. crispus*, and *F. sexcostatus*, were compared with each other, it would be impossible to separate them into distinct species.

The specimens present all the appearance of genuine fossils of the Coralline Crag, though from their locality (see footnote, p. 3) a question may attach as to this.

Fusus ? obscurus, *S. Wood*. 2nd Sup., Tab. I, fig. 12 *a, b*.

Axis, ⅞ths of an inch.
Locality. Cor. Crag, ? Boyton.

A single specimen, to which I have given the indefinite or undefined generic name of *Fusus*, was kindly sent to me by Mr. R. Bell. Although the shell is perfect it is decorticated throughout, and it is impossible to say whether it was, in its perfect condition, striated or not; but in its present state I cannot discover any trace of striæ upon it. I give it therefore under the above name from its uncertain characters.

Fusus? exacutus, *S. Wood*. 2nd Sup., Tab. II, fig. 18.

Locality. Cor. Crag, Sutton.

Our present figure represents only a fragment of a shell which has been in my Cabinet for many years. It was found by myself at Sutton in the upper portion of the Coralline Crag, and I have kept it hitherto unfigured in the hope of a better specimen turning up. On the left or columella side of the aperture is the impression of what appears to have been that of the fleshy lobe of the animal, but it is not represented in the engraving. The large opening in the outer lip is too low for a sinus, and is, I believe, simply a fracture. I think the specimen belongs to the genus *Fusus* and not to *Pleurotoma*. I now figure it because at my advanced age I must relinquish the hope of seeing a more perfect specimen.

Fusus nodifer, *A. Bell*, MS.   2nd Sup., Tab. III, fig. 4 *a, b*.

Locality.   Red Crag, Waldringfield.

The specimen here represented is from the Cabinet of Dr. Reed, and was obtained by Mr. Alf. Bell, who had affixed to it the above name and the following description :—" Shell fusiform, volutions 5, convex, with a ridge at the section, and eight or nine rounded ribs covered with coarse spiral striæ." The specimen is much rubbed and worn, and it is doubtless derived from an older formation.

At p. 117 of my first Supplement reference is made to the name of *Fusus despectus*, Linn., which has been given in the list to the paper of Mr. Prestwich as a species new to the Crag, and also in Mr. A. Bell's list of Crag shells.   I have made every endeavour to ascertain where the specimens are upon which this name has been founded, but without success.   In my large series of the abundant Red Crag shell, *antiquus*, nearly every form of exterior ornament, from the very finely striated specimens to such as are ornamented with large and prominent spiral ridges, like those upon *F. despectus* ('Ency. Meth.,' pl. 426) may be seen ; but this latter shell in the recent state has apparently a slightly curved outer lip, and this variety I have not seen from the Crag.   *Fusus tornatus*, Gould, is another proximate form, but in this the canal seems to be a little more oblique than in that of the Crag shell, and if these characters be the only differences all three might, I think, be united as varieties of one species.

Mr. Jas. Reeve has recently sent to me a specimen from the Norwich Museum which, he says, was found at Bramerton ; the name of *Fusus antiquus* accompanied the shell, and in this I believe he is perfectly right.   It appears to have lost the whole, or very nearly so, of the thick outer layer of the original shell, and in its present state, it somewhat resembles what I have called *Trophon altus*, so much so that if it had been *entirely* denuded by the removal of the outer shell it could not have been recognised for what it really is.   So many specimens from the Crag have suffered more or less by the removal of either the outer layer of the shell, or partially so in the destruction of some of its ornamentation, that I mention this case as an instance of the liability to which palæontologists are sometimes misled, by such alterations in the condition of the shell into the adoption of new species or of new identifications.

A specimen also from Dr. Reed has recently been sent to me with a label on which is written " *Fusus antiquus*, L., Cor. Crag, Broom Pits, near Orford, from the upper beds."   This is nothing but a recent specimen filled with and partially stained on the surface by the Cor. Crag material.   I have not yet seen this species (*antiquus*) from the Cor. Crag.   The shell which I have figured as *Trophon elegans*, is in the list of Mr. Prestwich's paper, p. 492, called a variety of *antiquus* ; but so far from assenting to that reference, I rather believe the shell to be the type of a new Genus, as suggested by Mr. Charlesworth, who figured and described it in the 'Mag. Nat. Hist.,' vol. i, p. 219, fig. 23 ; as it has asmall apex, and a deposit of calcareous matter on the upper part of the left lip.[1]

MUREX REEDII, *S. Wood.* 2nd Sup., Tab. I, fig. 9 *a, b.*

MUREX REEDII, *S. Wood.* Quart. Journ. Geol. Soc., vol. xxxiii, p. 120, 1877.

Spec. Char. Testá fusiformi, crassa ; spirá elevata ; apice acutá, anfractibus septenis subangulatis ; varicibus tenuibus, sublamellosis, ultimo anfractu maximo ; aperturá ovatá, labro intùs incrassato dentato ; columellá incurvata.

Length, 1⅝ths inch.
Breadth, ⅞ths inch.
Locality. Cor. Crag ? Boyton.

A specimen is among the shells sent to me by Dr. Reed, and from the perfection in which it was found, I am enabled to make a fair comparison of it with other shells of this genus in similar condition. It has prominent varices, which are not much foliated. It somewhat resembles *M. tripartita,* but is more elongated, and differs from it in not having spiral striæ like that shell, or like the long known Crag shell *M. tortuosus,* J. Sow , which is covered with large and prominent spiral striæ or ridges.

The artist's representation (figs. 9 *a, b,* of Tab. 1) might raise the idea that our present shell was obscurely striated, but I can detect no striation, though there are some faint transverse marks between one pair of varices, and as the shell is in such a fresh and unworn state it may be safely said that it never possessed striations. I have endeavoured by sending accurate drawings of the shell to Dr. Nyst, and several other Belgian conchologists, to ascertain whether anything like it was known from the Belgian beds ; but they all assure me that they know of nothing like it. The canal and mouth are slightly oblique (a feature which the artist has failed in the engraving to catch), and there are six varices on

[1] I may mention here that a dead and bleached specimen of *Conus tulipa* was once showed to me, and said to have been found in the Cor. Crag at Ramsholt ; and I have also seen a very pretty (fabricated) shell as a Red Crag fossil from Walton-on-the-Naze. This was a thick specimen of *Buc. Dalei,* beautifully ornamented with elevated ridges in a *Harpa*-like fashion, and executed in a very skilful manner, but the artist had left unobliterated a few small marks of his graving tool. These specimens are probably still in existence, and I mention them here like that of *Fusus antiquus* from Orford by way only of caution.

the body whorl and upon the preceding volution.   The apex probably was sharp, but the specimen is there slightly broken.   The shell is not quite so robust in proportion to its length as the artist has represented it.   It somewhat resembles *M. Haidingeri*, from the Vienna beds shown in Tab. 23 of Dr. Hornes' work ; but his figure differs from our present shell in having no denticulations on the outer lip, and in having the varices strongly continued down the canal.

In consequence of the unsatisfactory representation to which I have referred, I annex a cut made from a drawing which shows the characters of the shell more accurately.

The appearance of the specimen is not at all suggestive of its being a derivative ; and though obnoxious to the uncertainty which I have before (p. 3) mentioned as attaching to the specimen from Boyton, the specimen presents altogether the appearance of a genuine fossil of the Coralline Crag.

MUREX PSEUDO-NYSTII, *S. Wood*.   Tab. I, fig. 8 *a, b*.

M. Testá elongato-fusiformi, crassá ; spirá elevatá, anfractibus septenis, convexis ; supernè subangulatis, spiralite rlatè striatis ; varicatis, varicibus, 7—10, tenuibus, lamellosis, compressis ; ultimo anfractu equaliter longiore ; aperturá ovatá, canaliculata, canali attenuato, labro intus pauci denticulato.

Axis, 1⅛th of an inch.
Locality.   Cor. Crag? Boyton.

A perfect specimen as above represented has been sent to me by Dr. Reed, and so far as I am able to ascertain it appears to be specifically distinct from any previously described species.   The shell may be described as elongately fusiform, with seven or eight convex volutions, the upper part of these somewhat depressed, giving a slight shoulder to the volutions ; coarsely striated in a spiral direction, but above the shoulder these striæ do not extend : the apex was probably sharp and acute, but it is slightly broken ; aperture small and ovate, and the outer lip extremely thick ; and on which there were two prominent denticles, and one nearly obsolete on the lower part of the inner lip ; it has a long canal, slightly curved, and open.   The first two volutions appear to be smooth or destitute of marking either spirally or longitudinally.

I have compared it with specimens of Von Könen's species *Nystii*, kindly sent me by Dr. Nyst, and with others from Edeghem, in Belgium,[1] sent me by M. Rutot, and although it approaches that shell in several respects, it does not do so sufficiently to justify any identity with it.   Nevertheless, to indicate its affinity I have assigned it the

[1] This deposit of Edeghem has hitherto been regarded as miocene, but it is placed by M. E. Vanden Broeck with that of Kiel and some other localities near Antwerp as oldest Pliocene "Esquisse Géologique et Paléontologique des depots Pliocénes des environs d'Anvers," p. 35.

above name. *Nystii* is a less tapering shell, and possesses only half the number of varices, and these more thick and prominent than those of our present shell.

The same remark in reference to the genuineness of the shell as a species of the Coralline Crag, which I have made in the case of the last described species (*Reedii*), applies to the present case.

Two imperfect specimens, or rather the larger portion of some small species belonging to this genus, were found by myself many years ago in the Cor. Crag of Sutton, and were retained in the hope that something better would turn up to enable me correctly to describe them, or to refer to some previously described species. These are shown in figs. 7 *a*, *b* of Tab. I, and exhibit the last volution with the aperture and its straight canal perfect; and as these constitute the principal portion of the shell, a fair idea of it may be thus formed. The specimens very much resemble *Murex Canhami*, figured in No. 14 of Tab. VII of my first supplement in their coarse spiral striations, but they have not the prominent points or shoulders to the varices which that shell possesses, and their canals are straight and narrower than that of *Canhami*. In their imperfect state I have here called them provisionally *Murex recticanalis*.

MUREX CROWFOOTII, *S. Wood.* 2nd Sup., Tab. I, fig. 15.

Locality. Cor. Crag, ? Boyton.

The specimen figured is imperfect, as shown by the fragment of the last whorl which remains adherent to the preceding one, but in other respects is in finely preserved condition. The cross striation, which is very thick and strong, resembles that in *M. tortuosus*, but the form of the shell is much less elongated, and the number of distinct whorls preserved would seem to indicate that, when perfect, the specimen could be only that of a much smaller shell than *tortuosus*. As it was placed in my hands by Mr. W. M. Crowfoot, to whom it belongs, I have given it under the name of *Crowfootii*, which will also serve to indicate the ownership of the specimen, for comparison in the event of any one more perfect turning up. I am informed by Mr. Robert Bell that he has obtained many specimens of *M. tortuosus* from the Coralline Crag, which confirms my belief that this species which was long known from the Red Crag only, is merely present as a derivative in that formation.

TRITON CONNECTENS ? *S. Wood.* 2nd Sup., Tab. I, fig. 14 *a*, *b*.

TRITON HEPTAGONUS, *S. Wood.* Crag Moll., vol. i, p. 41, tab. iv, fig. 8, 1848.
    „   CONNECTENS, *id.* Supplement to Crag Moll., p. 30, 1872.

Axis, 1 inch.
Locality. Red Crag, Waldringfield.
A specimen of this genus has been sent to me by Mr. R. Bell, which he says is from

Waldringfield, that receptacle for so many derivatives; and as this shell is very rare to my researches, and the present specimen presents differences from the one previously represented, I have had it figured as above.   It is doubtless derivative.

RANELLA ? ANGLICA, *A. Bell.*   2nd Sup., Tab. III, fig. 3.

RANELLA ANGLICA. *A. Bell.*   Ann. and Mag. Nat. Hist., May, 1871.

Spec. Char.   " Shell small; whorls 3, 4 (apex wanting), convex, with coarse elevated ridges on the bottom whorl, crossing the periodic growths (which are very distinct), and extending to the mouth, becoming very marked at the base; mouth angulated above, outer lip spreading towards the base, where it is sharply angulated by one of the ridges; pillar reflected; canal rather open; umbilical chink small."—*A. Bell.*

Length, $\frac{6}{10}$ths of an inch.

Locality.   Red Crag, Waldringfield.

The only specimen of this shell which has been obtained, so far as I know, is the one now figured.   It is from Dr. Reed's collection, and was described as above by Mr. A. Bell.   It is not in a perfect condition, and I am doubtful of the correctness of the assignment, but have thought it best to have it figured, and give it under Mr. A. Bell's name and description.   It is no doubt derived from some antecedent formation, and seems to me to resemble a good deal the imperfect specimen from the Cor. Crag, figured by me in Tab. II of my first Suppt., under the name *Murex corallinus.*   There are some spiral striæ or ridges on the base or lower part of the volution, but the specimen is too much mutilated on the spire to show whether it was covered entirely with striæ. There are three or four distinct denticles on the inside of the outer lip, as in *M. corallinus,* and a few coarse ridges on the outside of this outer lip, as if the spire had also been so covered.

PLEUROTOMA MORRENI, *De Konninck.*   Tab. II, fig. 6 *a, b.*

PLEUROTOMA MORRENI, *De Kon.*   Desc. Coq. Foss. de Basele, p. 21, pl. i, fig. 3, 1837.
        „   „   *Nyst.*   Coq. Foss. de Belg., p. 510, pl. xl, fig. 6 *a, b,* 1843.
        „   INTORTA (?), *Bellardi.*   Foss. del Piedm., p. 16, tav. i, fig. 13, 1847.

Axis, 1¼ inch.

Locality.   Red Crag, Waldringfield.

The specimen as above represented is from the Cabinet of Mr. Canham, who tells me he obtained it from the well-known phosphatic nodule pit at the above-named locality.

M. Nyst, as also M. de Konninck, appear to think the shell referred to is a species distinct from *Pl. intorta*, Broc.; and as the Belgian shell seems not to be rare, and to have been found in good preservation, probably they have good means for such determination. In 'Crag Moll.,' vol. i, tab. vi, fig. 4, I figured two specimens of which the smaller one may possibly be the same as our present shell, except that it is more elongated and has a less pointed termination, and as I am not imposing a new name I have thought it best to figure and describe our present shell which, however, much resembles fig. 13, tab i, of M. Bellardi's paper. This naturalist, however, seems to consider the shell so figured by him as only a variety of Brocchi's species.

The Waldringfield specimen is doubtless derivative, but from what formation it has come is, of course, conjectural. Considering, however, the close resemblance of the Cor. Crag shell which I have figured under the name *Fusus Waelii* to a shell which occurs at Baesele, (the locality from which De Konninck describes our present species,) it is quite possible that our present shell may be among the many yet unrecognised species of the Cor. Crag which by the destruction of this Crag have gone to fill that museum of derivatives which the Waldringfield Red Crag accumulation constitutes.

PLEUROTOMA CURTISTOMA ? *A. Bell.* 2nd Sup., Tab. II, fig. 9, *a, b*.

PLEUROTOMA CURTISTOMA, *A. Bell.* Ann. and Mag. Nat. Hist., 1871, p. 7.

Axis, 1 inch.

Locality. Cor. Crag ? Boyton.

The shell represented has been recently sent to me by Dr. Reed, and it was, he tells me, obtained from the above-named locality. In colour it resembles the Coralline Crag. From the description given by Mr. Bell I have referred it doubtfully to *curtistoma*, but I have not had for examination the specimen to which Mr. Bell assigned that name, which I believe has gone into the British Museum. He gives for it the locality Cor. Crag, Gedgrave. The shell now figured is closely connected with one that I figured in my first Suppt. under the name of *Pleurot. Bertrandi* (?), Addendum Plate, fig. 4, p. 179, but it has a smaller and shorter aperture. In Mr. Prestwich's List, p. 494, *Pl. curtistoma* is given as a variety of *Pleurot. attenuata*. I think, however, that our shell is distinct, as it is not attenuated and has a shorter aperture, but more and better specimens than I have seen will be necessary for certain determination.

PLEUROTOMA TERES ? *Forbes.*   2nd Sup., Tab. II, fig. 7, *a, b.*

> PLEUROTOMA TERES, *Forbes.*   Ann. and Mag. Nat. Hist., vol. xiv, p. 412, pl. x, fig. 3.
> MANGELIA TERES, *Forbes and Hanb.*   Brit. Moll., vol. iii, pl. cxiii, figs. 1, 2.
> DEFRANCIA TERES, *Jeff.*   Brit. Conch., vol. iv, p. 362, pl. lxxxviii, fig. 5.

Axis, $\frac{5}{16}$ths of an inch.

Locality.   Cor. Crag, Sutton.

A small and worn specimen was found by myself some years ago in the Cor. Crag of Sutton, which I have kept unfigured in the hope of obtaining another and better preserved specimen to assist in its correct determination, but without success.   I now give it as above, but with a mark of doubt; and it is evidently distinct from *tereoides,* ' Supplement to Crag Moll.,' Addendum Plate, fig. 3 *a, b.*   In the ' Crag Moll.,' vol. i, tab. vi, fig. 6, is figured a minute shell with a peculiar ornamentation on the young or upper volutions; this was called *Trophon paululum,* and considered as the young of a larger shell.   In Professor Prestwich's paper, ' Quart. Journ. Geol. Soc.,' vol. xxvii, p. 146, this is referred to *Pl. teres,* which probably it is (see 1st Supplement to Crag Moll.,' p. 27).   My present specimen is somewhat abraded, and shows more numerous and close spiral striæ than the recent *teres* usually presents.   These in my specimen are not carried over the ribs, but this may be due to obliteration from wear; the ribs also are more prominent than in the recent shell.   On the other hand, the form of the shell, and its deep and broad sinus, agree with the recent species.   The striæ on the lower whorls are rather more numerous than represented by the engraver.

PLEUROTOMA GRACILI-COSTATA, *S. Wood.*   2nd Sup., Tab. II, fig. 8.

Spec. Char.   *Testá ovato-fusiformi, ventricosá, brevispirá, acuminatá; anfractibus convexis, longitudinaliter et angustè costatis; transversim striatis; ultimo basi sulcato; columellá canalique brevi, contortis; aperturá ovatá.*

Axis, $\frac{5}{8}$ths of an inch.

Locality.   Cor. Crag, Sutton.

The specimen figured was found by myself many years ago, but from its peculiar appearance I postponed noticing it, hoping that something better might turn up to assist in its determination.   It occurred to me that the costæ or ribs which are formed by the periodical arrest of the outer lip during growth might have been originally round and hollow, and that the upper part of them had been decorticated, and a portion consisting of the two sides of the original ribs only left, the effect of which would be to show a number of thin sharp, instead of half that number of wide and blunt costæ. The apex is sharp, the three first volutions being without riblets, and the fourth volution has 4 or 5 rounded riblets, beyond which these riblets are double in number. My specimen is not sufficiently perfect to show if there have been any spiral striæ. The outer lip is much curved and there is a large deep sinus a little below the suture; the outer lip is also sharp, without any striæ or ridges on the inside of it. My specimen resembles the figures given by M. Nyst with the name of *Pleurot. acuticosta* ('Coq. foss. de Belg.,' p. 529, pl. 42, fig. 5), but that figure is indifferent, and the description is too short to supply the deficiency. *Pleurot. incrassata* from Touraine somewhat resembles our shell, but I have not a specimen for comparison. The above name is given provisionally.

PLEUROTOMA ICENORUM, *S. Wood.* 2nd Sup., Tab. III, fig. 8, *a, b*.

PLEUROTOMA ICENORUM, *S. Wood.* 1st Supplement Crag Moll., p. 35.

Locality. Cor. Crag near Orford.

There is so much doubt and difficulty about this shell that I find it necessary to give another figure of it, from a perfect specimen in my own cabinet. My shell has a row of nodules formed at the projecting portion of the outer lip, with a row of smaller nodules adjoining the suture; thus making two rows on all but the lower volution. The two apical whorls are quite smooth and without ornament, making the apex very obtuse; differing thereby from the representation of *Pl. coronata* of Bellardi. At the base there is an umbilicus caused by a slight obliquity of the volutions outwardly. Two specimens have been sent to me from Dr. Reed's collection with the name of *Pl. umbilicata*, A. Bell, which correspond with *Icenorum.* Our shell has unfortunately had several names. In Mr. Prestwich's list, p. 145, it is called *Pleurotoma galerita*, Phil. In Mr. Bell's List of the English Crags, p. 35, it is said to have been figured and named by Dr. von Könen as *Pl. Hosiusii* ('Mioc. Nord. Deutch. Moll.,' p. 105, taf. 2, fig. 12 *a, d*). These foreign species appear to me (judging from representations) to be different from our shell, which has an obtuse apex and an umbilicus, neither of which is possessed by them. The name of *Pl. semicolon*, given in Crag Moll., is also erroneous for the reasons mentioned in my first Supplement, p. 35. I would have adopted Mr. Bell's name of *umbilicata*, were it not that the shell to which I had previously assigned the name *Icenorum* is, in my opinion, the same species.

PLEUROTOMA SENILIS, *S. Wood.* 2nd Sup., Tab. III, fig. 2 *a, b.*

> PLEUROTOMA SENILIS, *S. Wood.* 1st Supplement, p. 42, tab. v, fig. 5.
> — ARCTICA?, *Adams.* — — p. 45, t. vi, fig. 9.
> — VIOLACEA, *M. & A.* — —

Locality. Red Crag, Sutton and Waldringfield.

The original specimen, figured in my first Supplement, was very much worn, but some better preserved specimens from the Red Crag have been obtained by Mr. Canham. That which I have now figured as 2 *b* was the most perfect, and has since been lost by him, but having while it was in my hands had a drawing made of it I am enabled to give the figure 2 *b* from this. The specimen figured in Tab. V of my first Suppl. was so much rubbed that some uncertainty attaches to its identification with the shells now figured, and under these circumstances it is our present shell that I desire to distinguish by the specific name of *senilis.* The fragment, No. 9, figured by me in Tab. VI of my first Suppl. under the name of *arctica,* seems to be one of a much worn specimen of the present species. They are all derivative in the Red Crag, but may, I think, not improbably have been derived from the Coralline, though nothing identical with them has yet been obtained from that formation. Under the circumstances explained above, I have removed the name of *P. violacea* from my Synoptical list.

PLEUROTOMA CATENATA, *A. Bell*, MS. 2nd Sup., Tab. III, fig. 5.

Axis, $\frac{9}{16}$ths of an inch.

Locality. Cor. Crag, Gedgrave.

The above figure is taken from a specimen in the Cabinet of Dr. Reed, which was obtained from the Cor. Crag by Mr. A. Bell, who had assigned it the above name in MS.

There is so much uncertainty attending many identifications of the species of this genus that I prefer giving the figure of the shell with Mr. A. Bell's assignment of it to expressing any opinion of my own about it.

The shell has eight volutions, very slightly convex, indeed nearly flattened; apex obtuse; embryonic whorls smooth; there are two rows of nodules, above which is the sinus and two smaller spirally nodulous lines; base of volution covered with prominent spiral lines; aperture ovate, with a canal of moderate length; the ornamentation, though not very well defined, appears to be its only distinction. The specimen figured is the only one which I have seen, and is by no means perfect.

PLEUROTOMA PANNUS, *Basterot.*   2nd Sup., Tab. III, fig. 6.

> PLEUROTOMA PANNUS, *Bast.*   Foss. du Sud-ouest de la France, p. 63.
> —   —   *Bellardi.*   Monog. delle Pleur., p. 27, tav. ii, fig. 2.
> —   DUMONTII, *Nyst.*   Belge Foss., p. 527, tab. xlii, fig. 4.

Spec. Char.   " *P. striis transversis, numerosis, minutis; striis incrementi decussatis.*"
—Bast.

Axis, ⅝ths of an inch.

Locality.   Cor. Crag, near Orford.

France: Saucats, Léognan, Dax.

Piedmont: Torino, Colli Tortonesi.

The specimen figured, which, however, is not quite perfect, was found near Orford by Dr. von Könen; and he has kindly sent me a specimen of the same species from Antwerp, which seems to correspond with our Crag shell.   Mr. A. Bell has introduced this name into his list of Coralline Crag shells, so that probably several other specimens may have been found, but that in my possession is the only one from the Crag that I have seen.   *Pl. catenata* of Mr. Bell strongly resembles it, and may be only a variety.

As with so many species of this variable genus, it is difficult to say whether the distinctive features which induce authors to make specific distinctions are in the present case constant; but the identification of the shell by so good a conchologist as Dr. von Könen, and the production by him of a specimen from Antwerp identical in character with our Cor. Crag specimen, gives me more confidence in the present identification than I should otherwise entertain.

The specimen figured as No. 1 of Tab. III was sent to me by Dr. Reed, and has a label attached with the name of *Borsonia prima* assigned by Mr. A. Bell, who gives for it the locality of "Red Crag, Waldringfield."   The shell looks like a deformity, and the ridge upon the columella accidental, as it is angular in form, and like a simple projection.   The shell is much abraded, and appears like a mutilated specimen of *Pleurotoma turrifera,* Nyst (*P. turricula* of Brocchi, figured in 'Crag Moll.,' Tab. VI).   As the name *Borsonia prima* may, perhaps, be introduced into lists of shells from the Crag, I think it best to give a figure of the specimen, to enable others to judge for themselves.

Borsonia prima, Bellardi, 'Monog. Pleurot. Foss.,' pl. iv, fig. 13, is, I think, a different shell altogether.

A specimen of *Pleurotoma* from Boyton has been very recently sent to me by

Mr. Cavell, of Saxmundham, which closely corresponds with *Pleurot. lævigata*, Phil., being quite destitute of costæ; but the shell cannot be described as "lævissima," as there are vestiges of spiral striæ remaining upon the Crag specimen. This is possibly the same as fig. 12, tab. vii, of 'Crag Moll.,' but it is distinct from fig. 15, tab. vi, of my first Supplement, which I think may be referred to *P. nebula* of Mont.

CANCELLARIA (ADMETE) AVARA ? *Say.*   2nd Sup., Tab. IV, fig. 5.

> COLUMBELLA AVARA, *Say.*   Gould. Invert. Massach., p. 313, fig. 197.
> CANCELLARIA AVARA, *A. Bell.*   Ann. and Mag. Nat. Hist., May, 1871.

Axis, ½ an inch.
Locality.   Red Crag, Waldringfield.

This is another imperfect and much worn specimen from Dr. Reed's Collection, but as it has been published by Mr. Bell in his list of Crag shells as a species of that forma-tion, I have had it figured as above. I am unable to give a full description of the specimen from its mutilated condition, but it possesses several folds or small ridges upon the columella, from which, and its general form, it seems referable to that group of the Cancellariæ to which the subgeneric name *Admete* has been given, but beyond that I can express no opinion of its identity, and I give it under the name *Avara* solely on the authority of Mr. Bell. It appears to me like a derivative. I have a very imperfect specimen of an elongated species of *Cancellaria* from the Coralline Crag, but it is too much mutilated to permit of its being even provisionally described. It does not, however, appear to have belonged to the same species as the above shell.

CANCELLARIA CRASSISTRIATA, *A. Bell*, MS.   Tab. III, fig. 16 *a*, *b*.

Axis, ½ an inch.
Locality.   Red Crag, Waldringfield

The figure is taken from one of two debauched specimens from the Red Crag of Waldringfield in Dr. Reed's Cabinet, which were obtained for him by Mr. A. Bell, and who has sent me the following rough note upon them:—"Specimens much worn and decorticated. There are about ten striæ on the body whorl, the most prominent being three on the most extended part of the volution, crossed by some broad obscure ribs; the outer lip is thickened inside at the top; inner lip reflected upon the pillar, showing in worn specimens an umbilical chink. The absence of teeth on the inner lip would place the shell in the section *Admete*." Whatever the specimens may prove to be, they are evidently derivative in the Red Crag.

CERITHIUM VARICULOSUM, *Nyst.*   Crag Moll., vol. i, p. 69, Tab. VIII, fig. 3; 2nd Sup., Tab. II, fig. 15.

Locality.   Red Crag, Walton Naze.

The figure given of this shell in the ' Crag Moll.' does not quite correctly represent the fossil found at Walton Naze, which in Prestwich's list is referred to *Cerithium reticulatum,* but which I believe is specifically distinct; the volutions of my fossil are more convex, and are not only destitute of thickened varices, but have a different ornamentation from the recent shell.   I have now figured a fragment found by myself at Walton Naze; and this has decidedly convex volutions, with three spiral and nodulous ridges, and a small one at the base; moreover, these spiral ridges are not equally distributed over the whorls, there being a wider space between the upper one and the suture, than there is between the others.   In *C. reticulatum* the volutions are nearly flat and have four equidistant nodulous striæ.   I have therefore retained the shell under the name originally given.

CERITHIUM GREENII? *Adams.*   2nd Sp., Tab. IV, fig. 16.

CERITHIUM GREENII, *Adams.*   Bost. Journ. Nat. Hist., vol. xi, p. 287, pl. iv, fig. 12.

Locality.   Chillesford Bed, Bramerton.

Two small but very perfect specimens of some species of the genus *Cerithium* have been sent to me by Mr. Reeve with the locality of " Upper bed at Bramerton."   I have a difficulty in referring them to anything previously described, and have therefore given them provisionally the above name.   The shell to which they present the nearest approach is *Cerithium Greenii,* C. B. Adams, figured and described by Gould (' Invert. Mass.,' p. 279, fig. 184), but I have not the recent shell to compare with it.   In ' Brit. Conch.,' vol. iv, p. 267, it is said that *C. Greenii* is the same as *Cerithiopsis tubercularis,* but my shell does not correspond with anything that I have seen of this very variable species.   It does not seem possible that it can be the young of *C. tricinctum,* though it does not exceed in length $\frac{3}{16}$ths of an inch, for it has seven volutions, which is repugnant to its being the young of any species.   The base of our very perfect specimens is quite free from striæ or markings of any kind, and the volutions, which have three nodules, are separated by a deep suture, the two forming the apex being smooth.   If the shell should prove distinct from *Greenii* the name *Reevii* might be assigned to it, as the specimen was found and sent to me by Mr. Reeve, of the Norwich Museum.

CHEMNITZIA INTERNODULA? *S. Wood.* *Var.* ligata, 2nd Sup., Tab. II, fig. 11.

> CHEMNITZIA INTERNODULA, *S. Wood.* Crag Moll., vol. i, p. 81, tab. x, fig. 6 ; 1st Sup.
> Crag Moll., p. 60, for normal form.

Axis $\frac{6}{10}$ths of an inch.

Locality. Fluvio-marine Crag, Bramerton.

The specimen here represented is in the Norwich Museum, and was sent to me by its curator Mr. Reeve. As it seems to differ so materially in form from the numerous specimens and fragments of *internodula* that I have obtained from the Cor. Crag, I have here figured it in juxtaposition with a representation (fig. 12) of one of my specimens from the Cor. Crag of Sutton. It may have been affected, like the *Littorinæ*, &c., by the brackish water, and consequently have much altered its normal form. If it be of the same species I would call it *Chemn. internodula*, var. *ligata;* and the latter might be adopted for its specific designation if,the shell should prove to be specifically distinct. The only difference, however, that I can see is that the Norwich Crag shell is much less slender, the internodulation being the same. Mr. Crowfoot has sent me several specimens of this species from the Crag found in the Beccles Waterworks Well, which corresponds with the Fluvio-marine of Bramerton. These, though rather more slender than the variety figured above, are yet nearer to it than to the usual Coralline to Crag form.

CHEMNITZIA SENISTRIATA, *S. Wood.* 2nd Sup., Tab. II, fig. 20.

> CHEMNITZIA SENISTRIATA, *S. Wood.* Quart. Journ. Geol. Soc., vol. xxiii, p. 120, 1877.

Spec. Char. Testá angustá, subulatá, elongatá, apice obtusá; anfractibus 8—9, *convexiusculis, spiraliter sulcatis, vel striatis; striæ senis, latis, depressis; aperturá subquadrangulatá; columellá rectá, simplici; labro intus lævigato.*

Axis $\frac{1}{4}$ of an inch.

Locality. Cor. Crag, Sutton.

This is the shell mentioned by me in the 'Crag Moll.,' vol. i, p. 84, as a var. of *similis* with spiral striæ, but no costæ. I now consider it as distinct and figure it under the above name. It approaches a shell called *Scalaria quadristrata* by Dr. Speyer ('Die Conch. der. Casseler. Tert.,' p. 181, tab. xxiv, figs. 7, 8), but the aperture of my shell is of a different form to the one there represented, and it has more numerous striæ than that species. The striæ upon the specimen now figured are six in number, broad and rather flat, separated by a narrow line, and the volutions are very slightly convex.

Chemnitzia similis ('Crag Moll.,' vol. i, p. 84, tab. x, fig. 11) strongly resembles the representations of a shell called *Scalaria?* (*Pyrgiscus*) *Leunisii*, Phil., from the upper oligocene given in Speyer's work, ("Die conch. der Casseler Tertiärbildungen." p. 180, tab. xxiv, figs. 10—12), but I have not been able to compare my shell with the original of this. The apex of my shell is obtuse or slightly reversed as in the shell represented by Dr. Speyer, and has ten volutions, with 12—17 upright or slightly sloping costulæ, traversed by six or seven spiral lines. The Crag shell, *similis*, though abundant, is seldom in perfection (the surface being often worn down or decorticated), and it is rather more cylindrical than the German species represented by Dr. Speyer.

SCALARIA TORULOSA, *Brocchi*. 2nd Sup., Tab. II, fig. 13.

> TURBO TORULOSUS, *Broc.* Conch. Foss. Subap., vol. ii, p. 377, tab. vii, fig. 4.
> SCALARIA TORULOSA, *Hornes.* Vienna Foss., p. 488, taf. xlvi, fig. 13 *a, b.*

Length 1 inch.
Breadth 4 lines.
Locality. Cor. Crag?, Boyton.
A single specimen of this species has been obligingly sent to me by Dr. Reed, and he tells me he obtained it from Mr. Charlesworth, who says it was turned out of the phosphatic nodule workings at the edge of the Butley river in the Parish of Boyton, to which I have already (p. 3, footnote) referred, and its reference to a particular division of the Crag is therefore somewhat uncertain, but unless it be a specimen from the nodule bed itself (in which case it would in all probability be derivative from a formation older than the Coralline Crag), it is to that division rather than the Red that I should refer it. I have little doubt but that it may safely be referred to the fossil called as above by Brocchi; it is also present in the Vienna beds. Our specimen appears to have been a good deal rubbed (which favours its derivative origin), and the fine striæ with which it was originally ornamented are nearly obliterated. I have also received from Mr. R. Bell a fragment of this species, with a notification that it came from the Red Crag of Waldringfield. This fragment is much mutilated and abraded, and evidently of derivative origin.

SCALARIA FIMBRIOSA, *S. Wood*. Crag. Moll., vol. i, p. 91, Tab. VIII, fig. 12; 2nd Sup., Tab. III, fig. 17 *a, b.*

Locality. Cor. Crag, near Orford.
The specimen now figured presents some differences from that figured in Tab. viii

of the first volume of the 'Crag Mollusca,' in having the varices closer, and a more distinct ridge round the base of the lower whorl, which I have endeavoured to show by fig. 17 *b*, Tab. III. It agrees closely with one, rather larger, sent me by M. Rutot, of Brussels, from Kiel, near Antwerp, a bed which has, until lately, been regarded as miocene, but which M. Vandenbroeck refers to the oldest pliocene,[1] and there can, I think, be no doubt of the identity of the two shells.

SCALARIA GENICULATA?, *Brocchi.* 2nd Sup., Tab. IV, fig. 11.

TURBO GENICULATUS, *Broc.* Conch. Foss. Subap., p. 659, t. xvi, fig. 1.

Locality. Cor. Crag, Sutton.

A small fragment of a species of the genus *Scalaria* is in my cabinet, which may possibly be referred as above, depending, as I am obliged to do, upon the figure and description by Brocchi. This seems to differ from all other species of the genus in being less strongly or coarsely costulated, and in having the spiral striæ broader and flatter, with a very narrow depression between them.

This is another instance in which I regret my inability to compare my own shell with a veritable specimen of the species to which I have referred it. Brocchi describes his species thus:—"T. subulata, anfractibus subrotundatis, costellis capillaribus, varice ad utrumque latus crassiore." This thickened rib is not visible in my fragment.

TURRITELLA (MESALIA) PFNEPOLARIS, *S. Wood.* 2nd Sup., Tab. II, fig. 14.

TURRITELLA PENEPOLARIS, *S. Wood.* Suppl. to Crag Moll., p. 53, t. iv, fig. 20.

T. Testá turritá, elongatá; apice acutá? anfractibus 10—12 convexiusculis striatis; suturá depressá; aperturá subovatá; columellá concaviusculá; labro tenui.

Axis 1 inch.

Locality. Cor. Crag, Suttton, and Cor. Crag?, Boyton.

The figures which I have previously been able to give of this shell have been those of fragments only, but I am now enabled to give a figure of the entire shell from one of two specimens sent me by Dr. Reed, which was obtained from the nodule workings at Boyton, but which, therefore, is of uncertain reference so far as its geological position is concerned, and may even be derivative, for it has been considerably abraded. It shows

---

[1] 'Esquisse Geologique et Paléontologique des Dépots Pliocènes des Environs d'Anvers,' p. 35, Brussels, 1876.

the form of the aperture, which more resembles that of those species from the Lower Tertiaries (such as *Turritella sulcata* and others) which were placed in a new genus proposed by Dr. Gray, 1840, and called *Mesalia*.

The engraver has in the figure shown the specimen in too perfect preservation, for the striations on the upper whorls are, in the specimen itself, obliterated, as are those also along the central portion of the lower whorls, and the aperture also is less perfect than represented.

TURRITELLA TAURINENSIS (?), *Michelotti*.   2nd Sup., Tab. II, fig. 19.

TURRITELLA TAURINENSIS, *Mich.*   Etud. Mioc. Inf., p. 84, pl. x, figs. 1, 2.

Locality.   Red Crag, Sutton.

This imperfect specimen of some species of the genus *Turritella* has been in my possession for some years.   The genus is one in which the determination of a species is most difficult from the great variation which individuals belonging undoubtedly to one species, such as those of *Turritella incrassata*, present, and out of which variation several species have been made.   The present specimen seems, however, to differ so much that I think it must be distinct from any of the forms of *incrassata*.   There is a difference in the thread-like arrangement of the striæ, and a greater convexity in the volutions, than in either *incrassata* or *terebra*.   A shell described by Dr. Speyer, under the name of *Turritella Geinitzii*, Cassel, ' Tert. Conch.,' p. 145, tab. xx, figs. 8—12, is not unlike the one now figured, and I have little doubt that our present specimen is a derivative in the Red Crag from some bed older than the Coralline Crag.   Figs. 16 and 17, Tab. II, represent varieties of *T. incrassata*, which may, I think, be referred to *T. acutangulata* and *T. subangulata*, Brocchi.

EULIMA NAUMANNI? *von Könen.*   2nd Sup., Tab. IV, fig. 22.

EULIMA NAUMANNI, *von Könen.*   Marine Mittel. Oligoc., t. xi, fig. 19.
—   —   *Speyer.*   Cassel. Tert. Conch., p. 202, taf. xxvi, figs. 12, 13, *a, b.*

Axis $\frac{3}{16}$ths of an inch.
Locality.   Cor. Crag, Sutton.

A single specimen in my cabinet differs so much from any of the species of *Eulima* known from the Crag that I have referred it provisionally as above, depending upon the representation of the species given in the works of Speyer, and von Könen.   So many so-called species in this genus present such trifling differences that before a correct determination can be made it will be necessary closely to compare the specimens themselves, which, in the present case, I have not been able to do. Our present shell corresponds with the size and form of the figure given by Dr. von Könen, but not quite so much so with the figure by Dr. Speyer, who refers his shell to Dr. von Könen's species. Dr. Speyer's figure, however, shows an obsolete keel (or the vestige of a keel) at the base of the volution, which is not visible in my specimen, nor in von Konen's figure. My specimen seems to have had a very slight curvature at the lower part of the outer lip, but as it is not quite perfect this is obscure. The apex is rather obtuse, and the volutions, of which there are 7—8, are very slightly convex, giving a depression, or great distinctness to the suture.

EULIMA HEBE, *Semper*.   2nd Sup., Tab. IV, fig. 18.

> EULIMA HEBE, *Semper*.   Palæont. Unters., s. i, 171 (*fide* Speyer).
> —   —   *Speyer*.   Cassil. Tert. Conch., p. 203, taf. xxvii, fig. 2.

Locality.   Cor. Crag, Sutton.
        Germany : Ober-Oligocene, Nieder-Kaufungen.
The specimen figured is the only one which I have seen, and was found by myself in the Cor. Crag of Sutton. Having now been enabled to compare it with specimens from the German beds, I can assign it as above.

EULIMA ROBUSTA, *A. Bell*, MS.   2nd Sup., Tab. IV, fig. 17.

Axis, $\frac{1}{2}$ an inch.
Locality.   Red Crag, Waldringfield.
This shell, from Dr. Reed's Cabinet, with the above name given to it by Mr. A. Bell, has recently been put into my hands. It somewhat resembles *E. acicula* of Sandberger, figured and described by Dr. Speyer, 'Cass. Tert. Conch.,' p. 205, tab. xxvii, fig. 4, but has apparently fewer and more convex volutions, and is not so elongate and tapering as that species. The apex of our specimen is broken, and the outer lip is nearly straight, like that of *Eul. intermedia*, but it differs from that species in the convexity of the volution. It is doubtless derivative in the Red Crag.

The shell figured in my 1st Supplement (tab. iv, fig. 25) as *E. stenostoma*, Jeff., has since been so injured as to be unrecognisable, so that I am doubtful of its correct assignment, and whether it may not be the shell given above under the name of *E. Hebe*, Semper.

On the other hand, I have specimens from the Coralline Crag of *Eulima* differing from *E. subulata* in the possession of a curved lip, which appears to be the only distinction from that shell upon which d'Orbigny's species of *subula* is founded. With this, and omitting, for the reason just given, *stenostoma* from the category, the following ten species of what I refer to the genus *Eulima*, with the exception of the derived *robusta*, have formed part of the Crag fauna, one of them, the doubtful *similis*, belonging to the newer or Red division only.

It must be confessed that some of these species are separated upon distinctions such as in more variable genera are considered only of varietal importance. Continental conchologists seem to consider the form of the outer lip as a good auxiliary character for separation, but I am unable to say if this be one on which a safe reliance can be placed. Shells of this genus are of a porcellanous structure and opaque, the lines of increase being invisible.

1. Eulima polita, *Linn.* Crag Moll., vol. i, p. 96, tab. xix, fig. 1 *b.* Curved outer lip.
2. — intermedia, *Cantraine.* Crag Moll., vol. i, p. 96, tab. xix, fig. 1 *a.* Lip nearly straight.
3. — subulata, *Donovan.* Crag Moll., vol. i, p. 96, tab. xix, fig. 3. Straight outer lip.
4. — subula, *D'Orbigny.* Prodrom., iii, p. 34, No. 478. Curved outer lip.
5. — bilineata, *Alder.* Sup. Crag Moll., p. 66. Spirally coloured.
6. — similis ?, *D'Orb.* Sup., Crag Moll., p. 65, tab. vii, fig. 6. Spire inflected.
7. — glabella, *S. Wood.* Crag Moll., p. 98, tab. xix, fig. 2. Apex obtuse.
8. — Hebe, *Semper.* 2nd Sup., Tab. IV, fig. 18. Elongated aperture.
9. — Naumanni ?, *von Könen.* 2nd Sup., Tab. IV, fig. 22.
10. — robusta, *A. Bell.* 2nd Sup., Tab. IV, fig. 17. Convex volution.

RISSOA COSTULATA, *Alder.* 2nd Sup., Tab. IV, fig. 23.

> RISSOA COSTULATA, *Alder.* Mag. Nat. Hist., xiii, p. 324, pl. viii, figs. 8, 9.
> — — *Forb. and Hanl.* Brit. Moll., vol. iii, p. 103, pl. lxxvii, figs. 4, 5.
> — — *Jeffreys.* Brit. Conch., vol. iv, p. 35, pl. lxviii, fig. 1.

Locality. Cor. Crag, Sutton.

A single specimen has very recently come into my hands from Dr. Reed, with the above-named locality given to it by Mr. A. Bell. This resembles in form *Rissoa crassistriata* of 'Crag Moll.,' vol. i, tab. xi, fig. 13, but that shell has large and coarse spiral striæ, of which the present species is destitute.

RISSOA PARVA ?, *Da Costa.* 2nd Sup., Tab. IV, fig. 21.

Locality. Cor. Crag, Sutton.

The specimen figured is from my own cabinet, and was found by myself. It appears to answer to this species, though from being unique and imperfect, I give it with doubt.

RISSOA RETICULATA, *Mont.*   2nd Sup., Tab. IV, fig. 19.

A specimen with this name has been sent to me by Dr. Reed, which seems to correspond with the recent British shell to which I have, as above referred, it. The shell so called in 'Crag Mol.,' vol. i, p. 163, tab. i, fig. 5, has been the subject of a criticism not easily to be understood (see 1st Suppt., p. 73). I have therefore had the present specimen figured, which is a more elongated form.

HYDROBIA OBTUSA, *Sandberger*.   2nd Sup., Tab. IV, fig. 7.

LITTORINELLA OBTUSA, *Sandb.*   Conch. de Mainz Tertiarb., s. 81, taf. 6, fig. 8 *a—c.*

Length 1 line.
Locality.   Fluviomarine Crag, Bramerton.
Several specimens of this little shell have been sent to me by Mr. Jas. Reeve, who tells me that he found them at Bramerton, and was doubtful about their correct assignment. The one figured is the longest of the series, and seems to approach very close to the figure of the shell given by Dr. Speyer from the middle oligocene of Germany, under the name of *Bithinia obtusa*, Sandberger; and as the specimens show the same thickened lip as does his figure, I have ventured to identify them with it. As the specimens are in good condition, and the allied species *subumbilicata, thermalis,* and *ventrosa,* which are abundant and in very perfect condition at Bramerton, are also figured by Dr. Speyer (under the name *B. acuta,* Drap.) from the same middle oligocene beds, I am disposed to regard the species now under description as having lived in the waters of the Crag Period equally with *subumbilicata ;* and not to be of the derivative origin of the shells described in the postscript.

NATICA (AMAUROPSIS) JAPONICA ?, *A. Adams*, M.S.   2nd Sup., Tab. III, fig. 11.

Axis $\frac{1}{4}$ of an inch.
Locality.   Red Crag, Butley.
A small specimen is among the shells sent to me by Dr. Reed, with the above name attached (by, I believe, Mr. A. Bell, who obtained it from Butley).
It is in good preservation and I have had it here figured, but whether it be the shell above

named I must leave for further observation and more specimens to determine. It much resembles a small form of *Natica helicoides* (*Islandica*, Gmel.), 'Crag Moll.,' vol. i, p. 145, tab. xvi, fig. 3, and may possibly be the young of that shell, though it seems to be more elongated, and to possess a more elevated spire and more pointed umbo; the present specimen is quite free from striæ of any kind, and it does not appear to have lost any of its outer coating, which is so common in specimens of *Naticæ* from that locality, and this is perhaps in favour of its being distinct. I have not been able to see the living shell to which Mr. Bell has referred it, which, on the label appended to our present specimen, is called "undescribed." The volutions in this specimen are convex, and between them is a deep and depressed suture, like that upon *helicoides*, but our present shell has a very distinct umbilicus. Mr. Bell tells me he has seen the young of *N. helicoides*, and that our present shell differs from it. I have put a mark of doubt against the present name, as I have not much confidence in the above assignment.

NATICA GROENLANDICA ?, *Beck.*, var. *declivis*. 2nd Sup., Tab. III, fig. 12 *a—b*; Crag Moll., vol. i, p. 146, Tab. XII, fig. 5; 1st Sup., p. 75.

Axis ⅞ths of an inch nearly.
Locality. Red Crag, Butley.

The shell now figured differs so materially from all the Crag *Naticæ* that I have been at a loss to what it should be referred. Its elevated spire almost brings it into what has been generically called *Amauropsis*, but as I believe it to be a true *Natica* I have preferred to give it here simply as a very abnormal form of some known species of that genus; and as *N. Groenlandica* seems to answer to it in respect of the more reliable characteristics upon which the species of *Natica* have been separated, and is withal a variable species, it is to this that I provisionally assign it as a variety (*declivis*). I am reluctant to assign new specific names on the evidence of a solitary specimen where the distinction of it from any other known form is not clear, but if further specimens of this shell should be found, then I think it might be regarded as a new species under the name *declivis*.

NATICA TRISERIATA ? *Say.* 2nd Sup., Tab. III, fig. 14, *a—b*.

NATICA TRISERIATA, *Say.* Journ. Acad. Nat. Sci., v. 211 (*fide* Gould).
— — *Gould.* Invert. Massachusetts, p. 233, fig. 165.

Axis 1 inch.
Locality. Red Crag, Butley.

The specimen figured seems to be intermediate between *Natica sordida* and *Natica Alderi*, approaching rather nearer to the latter than the former, but to neither does it strictly accord, having the form and nearly the size of *sordida*, but without its depression upon the upper portion of the volution. It is also rather more elongated than either, while the left lip is more extended than in *Alderi*, but rather less so than *N. sordida*. The shell is strong and nearly ovate, the contour showing but very little depression between the volutions, which slopes from the small and pointed apex. The exterior is smooth with simple lines of growth. As the specimens maintaining these characters are not rare I have ventured to refer them as above, though they bear a resemblance to *Natica hemiclausa*, a shell very abundant in the older part of the Red Crag at Walton Naze, but this latter has the umbilicus covered by the left lip in specimens that are full grown.

Naticæ are extremely abundant in the Butley bed, in association with the various peculiar and northern species of mollusca, which distinguish that newer portion of the Red Crag from the older or Walton portion, and their generally decorticated condition, in which the specimens which I refer to *triseriata* participate, increases the difficulties which attach to their specific separation.

I have not the recent species for comparison, and in making my reference to it my dependence is upon the figure and description given by Gould. The coloured markings which induced that author to give to it its name have disappeared in the Crag fossil, if they ever were present. There is also a resemblance between our fossil, and *Natica immaculata*, Totten, but this Mr. Jeffreys refers to *N. Alderi*, to which species I think the present fossil does not belong.

In 'Crag Moll.,' vol. i, p. 144, I said, when speaking of *Natica varians*, " It appears to be quite distinct from *Natica hemiclausa*, and it agrees in most of its characters with *N. varians* from Touraine." I am still of the same opinion. In Mr. Prestwich's List, p. 144, *N. varians* of the Cor. Crag is referred as a variety to *N. cirriformis*, but *N. cirriformis* is *there* referred to *N. sordida*. In Mr. A. Bell's List of the Lower English Crag, *N. varians* of the Crag is considered as *N. helicina*, Broc. The same shell is by M. Nyst figured as *Natica hemiclausa*, Sow. These conflicting opinions afford a proof of the perplexity in which those who study fossil mollusca become involved when occupied with this genus.

I have in Tab. III, fig. 7 *a—b*, given the representation of another specimen of this genus from the Coralline Crag near Orford, which is in Mr. Cavell's collection. This seems to differ materially from the shell which I have figured as *N. helicina* from the Red Crag of Walton Naze (' Sup. Crag Moll.,' p. 74, fig. 8 *a*, *b*), as it possesses a large and deep umbilicus, and although the front of the shell shows a depression at the suture, there is remaining a small portion of shelly matter, which if continuous would cover this deep suture entirely, and indicate that it possessed this covering feature, which is wanting in *N. helicina*.

Being a solitary specimen and surrounded by this uncertainty I have not ventured to assign it as a new species, preferring to give it as a variety, *heliciformis*, of *N. helicina;* but should more specimens occur maintaining its characters that varietal name might be assigned to it specifically.

In Tab. IV, fig. 12, of my first Supplement, is represented a specimen under the name of *N. proxima*, S. Wood, and at p. 74 of the same Supplement, the shell so represented is referred to the species figured in Tab. XVI of my original work under that name. As, however, the specimen in question does not show the depression on the upper part of the volution, and seems to be identical with the shell above given as *N. triseriata*, this reference was, I now consider, erroneous; and the figure should be regarded as one of the last-named species.

AMAURA HESTERNA, *S. Wood.*   Figured in the margin.

Axis.   ¼ of an inch.

Locality.   Crag, Boyton.

Spec. Char.   Testá turritá, elongato-conoideá, nitidá, glabrá; apice obtusá et depressá; anfractibus convexiusculis 5—6; suturis distinctis; aperturá brevi pyriformi: labro acuto simplici.

Mr. Robert Bell has sent me a specimen, but without a name, which he says came from Boyton, and which appears to belong to the same genus as the specimen figured in my first Supplement under the name of *Amaura candida*, Tab. I, fig. 3, from the Red Crag of Butley, and of which a very perfect specimen was also obtained by Mr. Crowfoot from the locality of Boyton.   This latter specimen, however, was stained with the Red Crag colour as much as was the Butley specimen, and undoubtedly belongs to the Red Crag.   The specimen I am now describing, however, though evidently of the same genus, is not only specifically different from *candida*, but is unstained with any red colour, for it is polished and nearly colourless.   It has the two apical volutions shallower and more depressed comparatively to the others, the suture distinct and somewhat deep, the aperture elongately ovate, terminating acutely at the body of the volution,

Amaura hesterna, *S. Wood*, enlarged ⁴⁄₁.

the outer lip sharp and simple, with a small but distinct umbilicus, and the body whorl occupies more than half of the entire shell.

This and *candida* are the only species of the genus at present known to me.   Their generic character is particularly indicated by the uppermost whorls that succeed the apex being unlike those which follow them, for instead of maintaining the proportions with which the shell commences to grow, the whorls increase in depth far beyond the proportions due to the increasing size of the animal, so that the angle of volution becomes greatly diminished.   In fact, the Mollusc appears to begin life under the form of *Natica,* and, after the growth of two whorls, to change its form so as to produce a shell quite unlike the oblate form of *Natica*, and of a more cylindrical shape.   Our present shell is much more tapering than *candida*, and it possesses also one more whorl than the Red Crag specimens of that species, though it has only half their linear dimensions.   It therefore seems to be a full-grown shell.

ADEORBIS ? NATICOIDES, *S. Wood*.   2nd Sup., Tab. III, fig. 13 *a, b*.

Diameter, $\frac{1}{10}$th of an inch.

Locality.   Cor. Crag, Sutton.

A small shell has been in my hands for many years, found by myself in the Cor. Crag of Sutton.   This has always much perplexed me, and it remained in my cabinet unfigured and undescribed from the idea that it might be the young or embryo condition of some larger species, and in the hope that I might obtain something further to assist in its correct determination.   Not having succeeded in this, I now figure the specimen as above. I have a large number of very small specimens of several species of *Natica*, and have broken up many of them with the expectation that I might produce something that would show a keel round the umbilicus similar to the one in my present specimen, but without success.   There is a large umbilicus in some species of *Natica*, but in none can I find any ridge around this great opening such as the shell now figured presents.   Two very anomalous shells, having large umbilical openings surrounded by a keel, have been figured by the late M. Deshayes, viz. *Lacuna mirabilis*. ' An. du Bas. de Par.,' vol. ii, p. 372, Pl. XVIII, figs. 1—4, and *Sigaretus problematicus*, vol. iii, p. 90, Pl. LXIV, figs. 7—9 ; but neither of these correspond to our present specimen.   There is also the living British species, *Lacuna pallidula*, which possesses a somewhat similar keel round an open umbilicus ; but our shell has a distinct ridge or keel *within* the umbilical aperture, of which no species of *Lacuna* that I have examined shows any trace.

Delphinula trigonostoma, ' Bast. Bord. foss.,' p. 28, Pl. IV, fig. 10 (which I had given as a synonym to *Adeorbis subcarinata*, but I believe erroneously), is perhaps the nearest approach to my shell.   I feel that the reference of the shell is very doubtful, but I give it to draw the attention of collectors.

TROCHUS ZIZIPHINUS, *Linn*.   2nd Sup., Tab. IV, fig. 20 ; Crag Moll., vol. i, p. 124, Tab. XIII, fig 9 ; 1st Sup., p. 81.

Dimensions.   *Height*, $\frac{1}{13}$th inch.

Breadth, $\frac{11}{16}$th inch.

Locality.   Cor. Crag, Sutton.

The present shell is from the collection of Mr. Canham, who tells me he procured it from the lower portion of the Cor. Crag at Sutton, and I have figured it in consequence of its unusual size. This shell was originally figured in Min. Conch. under the name of *T. lævigatus*, and figured under that name by Nyst from the Belgian beds. In my catalogue (1842) I called it *pseudo-ziziphinus;* from its resemblance to the living *ziziphinus,* and in the first vol. of Crag Moll. gave it as identical with that shell. It appears to be identical in ornament (though not in form, being less tapering), with a specimen from the Sicilian beds in my cabinet. This is probably the same as the shell living in the Mediterranean called *conulus.* I have many Crag specimens, smaller than the one figured, in which the exterior with its ornamentation is in perfection; and this so agrees with that in *conulus,* that if our Crag shell called *ziziphinus* be only one of the living varieties of that species, I think *conulus* and *ziziphinus* should be united.

Assiminia Grayana? *Leach.* 2nd Sup., Tab. III, fig. 18 *a, b.*

> Assiminia Grayana, *Leach.* Fleming's Brit. Anim., p. 275.
> — — *Forb. & Hanl.* Brit. Mollusca, vol. iii, p. 70, pl. lxxi, figs. 3, 4.
> — — *Jeffreys.* Brit. Conch., vol. v, p. 99.

Locality. Fluvio-marine Crag, Bramerton.

Two specimens have been sent to me by Mr. J. Reeve as from the "*Scrobicularia* bed at Bramerton,"[1] having been thought by him to be something different from *Hydrobia ventrosa.* One of these two I have here had represented, and I have referred it with some doubt as above, as it does not strictly accord with the living shell, which is obscurely angulated at the base of the last volution, like the shell of *Hydrobia ulvæ,* whereas in our present specimens the base is rounded. It differs materially from any specimen of *ventrosa* that I have seen, and has not the depressed or deep suture of *Bythinia Leachii.* In form it seems intermediate between *B. tentaculata* and *H. ventrosa.*

The shells at Bramerton being not unfrequently so distorted as to be scarcely recognisable for the species, or even genus, to which they belong, it is possible that the specimens in question are cases of this kind, so that I make the present reference with all reserve.

---

[1] This Scrobicularia bed at Bramerton appears to intervene between the few feet of specially Fluvio-marine Crag ($\overline{4}$ of sect. xvi of the Introduction to my first 'Supplement') which rests on the chalk and the Chillesford bed (5' of that section), thus answering exactly to the Scrobicularia beds at Butley, (4''' of sect. xvii of the same Introduction) to which the fourth column of the synoptical list refers.

Valvata cristata, *Müller.*   2nd Sup., Tab. IV, fig. 8 *a, b.*

> Valvata cristata, *Müll.*   Hist. Verm., pt. ii, p. 198.

Locality.   Fluvio-marine Crag, Bramerton.

This shell is abundant in the Freshwater deposits of Stutton, Grays, and Clacton, but I have only met with the one now figured from the Fluvio-marine Crag.

Valvata piscinalis.   2nd Sup., Tab. IV, fig. 9.

Locality.   Fluvio-marine Crag, Bramerton.

This is also very abundant in the same Freshwater deposits, but it is very rare in the Fluvio-marine Crag; it so closely resembles *Margarita helicina* that it is very difficult to distinguish the difference, and scarcely possible, except with perfect specimens; and I am doubtful whether a specimen found by Mr. Harmer at March, given by me at p. 121 of Vol. XXIII of the 'Quart. Journ. Geol. Soc.,' as *Trochus helicinus*, may not be merely *Valvata piscinalis*, since Freshwater shells occasionally occur in the March gravel.

The figure previously given of *V. piscinalis*, 'Crag Moll.,' Tab. XII, fig. 3, represents the depressed form, and I have given the more elevated one, which, when first discovered, was considered as a distinct species, and called *antiqua*.

The reference of *Margarita helicina* to the Coralline Crag made in my Catalogue of 1842 was an error.

Limnæa auricularia, *Linné.*   2nd Sup., Tab. IV, fig. 3 *a.*

> Helix auricularia, *Linn.*   Syst. Nat., edit. 12, p. 1249.
> Limnæa   —   *Jeffreys.*   Brit. Conch., vol. i, p. 108, pl. vii, fig. 4.
> Limnæus auricularius, var. acutus, *Forb. & Hanl.*   Vol. iv, p. 171, pl. cxxiii, fig. 2.

Locality.   Fluvio-marine Crag, Bramerton.

A single specimen, as above represented, has been sent to me by Mr. Reeve, and it is the first instance that I have met with of this species having been found in the Crag. It is, however, present in most of our newer Pliocene Freshwater beds, as may be seen in my List, 'Crag Moll.,' vol. ii, p. 307. Dr. Jeffreys gives three varieties to this species, our shell agreeing best with the one he first gave as distinct (*Limnæus acutus* in 'Linn. Trans.,' xvi, p. 373), but which he afterwards reduced to a variety. Our fig. 3 *b* was made from a recent specimen by mistake.

LIMNÆA PALUSTRIS, *Müller.*   2nd Sup., Tab. IV, fig. 2 *a, b.*

BUCCINUM PALUSTRE, *Müll.*   Verm. Tert. et Fluv., vol. ii, p. 131.

Locality.   Fluvio-marine Crag, Bramerton.
The shell figured and described in 'Crag. Moll.,' vol. i, p. 7, Tab. I, fig. 9, as *L. palustris* is, I think, there erroneously referred, as it more resembles the American species or variety called *elodes,* to which I would now refer it.   I have received from Mr. Reeve a specimen, of which the one above referred to is a representation, and which, I think, is the true form of *L. palustris.*

LIMNÆA PEREGRA, *Müller.*   2nd Sup., Tab. IV, fig. 4.

BUCCINUM PEREGRUM, *Müll.*   Verm. Hist., pt. xi, p. 130.

Locality.   Fluvio-marine Crag, Bramerton.
The shell now figured is the true form of the common variety of this species.   The one previously figured in 'Crag Moll.,' Tab. I, fig. 7, resembles the northern form called *L. Pingelii* by Möller, to which I will refer it.   Fig. 8 of Tab. I of 'Crag Moll.,' there called *L. truncatula* (?), corresponds with *L. Holbollii,* Möller, and I have not seen the true form of *truncatula* from any East Anglian bed.

PUPA EDENTULA, *Draparnaud.*   2nd Sup., Tab. IV, fig. 6.

PUPA EDENTULA, *Drap.*   Hist. Moll., p. 52, pl. iii, figs. 28, 29.

Locality.   Fluvio-marine Crag, Bramerton.
This has been obtained by Mr. Reeve, and he tells me it is from the " *Scrobicularia* bed " at that locality.[1]   The generic name of *Vertigo* is now given to this shell by some authors in consequence, it is said, of a difference in the animal, *Vertigo* having only two tentacles, while that of *Pupa* has four ; but there is nothing in the shell to denote a generic difference, and I have therefore retained its original name.   Our present shell is not rare in the newer Pliocene Freshwater beds, but it has not been hitherto given as a Crag shell, so far as I am aware.

[1] See note, p. 35.

MELAMPUS FUSIFORMIS, *S. Wood*, var. ELONGATUS.   2nd Sup., Tab. III, fig. 15; Crag
                                             Moll., vol. i, p. 12, Tab. I, fig. 14; and 1st Sup.,
                                             p. 3, Tab. I, fig. 1.

Locality.   Red Crag, Waldringfield.

The above specimen was obtained by Mr. Canham, and is perfect, except a slight
fracture in the back, which, however, is no injury to the shape of the shell.   It is more
elongated than any form of the genus that I am acquainted with, but, unfortunately, the
artist has not represented this character sufficiently in the present figure, which can
scarcely be distinguished from the original *fusiformis.*

BULIMUS LUBRICUS, *Müller.*   2nd Sup., Tab. IV, fig. 10 ; 1st Sup., p. 187.

> HELIX LUBRICA, *Müll.*   Hist. Verm., pt. xi, p. 104.
> ZUA LUBRICA, *Forb. & Hanl.*   Brit. Moll., vol. iv, p. 125, pl. cxxv, fig. 8.
> COCHLICOPA LUBRICA, *Jeff.*   Brit. Conch., vol. i, p. 292, pl. xviii, fig. 2.

Locality.   Red Crag, Butley.

The specimen figured is that referred to in my first 'Supplement' as found by Mr.
Canham, in the Crag of Butley, and although it is not uncommon in the Freshwater
deposits of Stutton, Clacton, Grays, and Copford, it is the first and only one that I have
seen from the Crag; I have therefore had it figured.   This shell has received several
generic names, but the above having been previously used in my list of the Land
and Freshwater shells in my second volume of the 'Crag Moll.,' I have not thought it
necessary to alter it here.

POSTSCRIPT.

DURING the progress of the foregoing through the press Mr. Jas. Reeve, of the
Norwich Museum, was good enough to send me a quantity of small shells, which he had
extracted from the sand of the Bramerton Crag Pit.   These consisted for the most part of
specimens of species already figured and described, but among them were two or three
which appear to me to be quite new to the Crag, if not, indeed, undescribed from any
formation.   These specimens are all more or less worn and imperfect, a character which
is not usual with the specimens of species belonging to any horizon of the Crag in
Norfolk ; and I feel little doubt that they are not shells which lived in the Crag waters,

but are derivatives from some other formation. As they approach species figured in Dr. Speyer's work from the Oligocene of Cassel, in Germany, nearer than they do to any others that I can find figured and described, I suspect that they have been introduced from some Upper Eocene or Oligocene formation in North-Eastern Norfolk, through which a stream flowed which discharged into the estuary of the Fluvio-marine Crag. The probability of such a thing is strengthened by the circumstance that the chalk disappears below the water-line of the country immediately east of the Bramerton Crag Pit, and by the Lower Eocene having been pierced at Yarmouth and found to extend to a depth of 526 feet below the sea level.[1]

The specimens in question comprise—

1. CERITHIUM DERIVATUM, *S. Wood*. Figured in margin.

Locality. Fluvio-marine Crag, Bramerton.

Two specimens of this species were among the shells sent by Mr. Reeve. One of these was so much worn and mutilated as to be recognisable with great difficulty, but the other, which is that represented in the zincograph, is in tolerable condition ; for though it has lost its apex, that is a thing not unfrequent with fossils of this genus, even where no suspicion of derivation attaches to them, and the surface is but little worn. It resembles the representation given by Dr. Speyer of *Cerithium Descoudresi*, from the Upper Oligocene, ' Cassel Tert. Conch.,' Taf. xx, fig. 2 *a, b ;* but his figure shows six distinct transverse or spiral lines, whereas the Bramerton specimen shows but four on the lower, and not so many on the upper whorls. With that distinction I have been unable to refer the specimen to Dr. Speyer's species, but as the number of transverse lines in this genus is not a constant character, it may, nevertheless, belong to it, and further specimens would determine that question. I have accordingly assigned to it provisionally the above name in order to distinguish its derivative origin. The specimens will be preserved in the Norwich Museum.

Cerithium derivatum, *S. Wood*, enlarged ⅔.

2. ODOSTOMIA? DERIVATA, *S. Wood*. Figured in margin.

Locality. Fluvio-marine Crag, Bramerton.

Several specimens of this shell were among the quantity already mentioned as **sent**

[1] Prestwich, in ' Quarterly Journal of the Geol. Soc.,' vol. xvi, p. 450.

me by Mr. Reeve ; but all of them were in a more or less mutilated condition.   One of the best preserved of them is represented in the accompanying zincograph.

Odostomia derivata, *S. Wood*, enlarged ¹²⁄₁.

The shell much resembles the figure of *Actæon lævisulcatus* of Sandberger (Nos. 4 and 5 of Taf. xxxiii of Dr. Speyer's ' Cassel Tert. Conch.'), a species of the Upper and Middle Oligocene of Germany ; but as neither the apex nor the mouth of any of the Bramerton specimens are perfect, I do not feel sufficient confidence in their identity to refer them to Sandberger's species, and have, therefore, given them under the above name provisionally.   The shading in the figure being effected by coarse lines gives the erroneous idea of the shell being covered with fine vertical lines.   It, however, possesses only the strong horizontal or spiral striæ shown in the figure.

Besides the above there was a single specimen of an *Odostomia*, which I am unable to refer to any Crag species or to any living British form ; but it is too much worn for me to venture to describe it as a new species.   It is about an eighth of an inch in length, and in its present state is free from striæ.   It is probably, like the foregoing, a derivative from some older formation.   There were also among the specimens fragments of the hinge portion of a small bivalve resembling the figure of *Siliquaria parva*, Speyer (' Ober. Oligocan Tert. Detmold,' p. 33, Taf. iv, fig. 2), but they are too imperfect for correct recognition.   There was also among them an imperfect specimen of a minute *Actæon*, which, I think, may be perhaps *A. Philippii*, Koch and Wiechmann (Die oberoligoc. Fau. des Sternberger Gesteins in Meckl.,' Abth. s. 7, Taf. i, fig. 3 *a—c*, represented by Speyer in Taf. xxxiv, fig. 1—3 of his work on the ' Cassel Tertiaries.')   It resembles that species in form ; and possessing four complete whorls, though only one eighth of an inch in length, it can hardly be the young of either of the Crag species *Noæ* and *tornatilis*.   As, however, I could not under a magnifyer detect the peculiar pitted marks which separate the striations in *A. Phillippii*, I have not ventured so to assign it.   Among the specimens there was also one of *Rissoa proxima*, Alder, which, though it has lost the upper whorls, is otherwise well preserved, and on the authority of it I have introduced that name into the Fluvio-marine Crag column of the synoptical list.   These specimens also will be preserved in the Norwich Museum.

BIVALVIA.

ANOMIA STRIATA, *S. Wood.* 2nd Sup., Tab. VI, fig. 3 *a—f;* Crag Moll., vol ii, p. 11, Tab. II, fig. 3; 1st Sup., p. 100.

Diameter. 1⅞ths of an inch.

Locality. Cor. Crag, Sutton and near Orford.

In my figure and description of this shell in the ' Crag. Moll.,' above referred to, the exterior only is represented. I now give, therefore, one of the interior of a specimen of similar magnitude, and also a separate fig. (3 *c*), representing the thickened portion of the lower valve, which resembles what I erroneously figured in ' Crag. Moll.' (vol. ii, Tab. XXXI, fig. 24), as possibly the internal shell of *Aplysia.* The lower valve of *Anomia* is very thin, except the ridge, which is represented in fig. 3 *d*, which, therefore, is the only part of this valve usually found; but fig. 3 *e* represents a perfect specimen of this valve, showing the opening for the byssus close to the connecting ligament.

Fig 3 *f* represents a small specimen of the upper valve from the Coralline Crag of Sutton, which shows that the shell in its young condition is perfectly free from striæ, these appearing when it is a little further advanced in life. This is the only specimen out of many hundreds that I have obtained from the Cor. Crag in which this feature is shown.

OSTREA UNGULATA, *Nyst.* 2nd Sup., Tab. V, fig. 7 *a, b;* Crag Moll., vol. ii, Tab. II, fig. 1 *a.*

OSTREA UNGULATA, *Nyst.*, var. A. Coq. Foss. Belg., pl. xxxiv, fig. 1.

Locality. Cor. Crag, Ramsholt.

I have here given another figure of the *Ostrea* occurring in the Coralline Crag, which was in the ' Crag Mollusca ' referred by me to *edulis*, and of which a specimen with the two valves united is represented in fig. 1 *a* of Tab II of vol. ii of that work. I am now inclined to think that this form is so far distinct from the common *edulis* that it should be separated from it.   The *O. edulis* of our coasts has the lower valve always more or less covered with imbricated radiations, of which the Cor. Crag shell is destitute, or on which, at least, they are obsolete or nearly invisible.   The *common* form of our edible Oyster has not come under my observation, either from the Coralline or from the Red Crag.   Figs. *a* and 2 *b* of Tab. II, ' Crag. Moll.,' may possibly be the immature state of *O. princeps*.   Our edible Oyster is described in ' Brit. Conch.,' vol. ii, p. 38, as having the " hinge-line narrow and nearly straight," " lateral edges (especially of the flat valve) finely crenulated or notched on the upper part ;" but the Cor. Crag shell is destitute of these, and the depression left by the connector is greatly incurved ; I have, in consequence, had the outside of the lower valve, as well as the place of the connector figured.

The Cor. Crag shell is very thick and ponderous ; and in that respect it resembles the more southern form of *edulis,* which Lamark described as a species under the name of *Ostrea hippopus.*   It, however, corresponds better with the Oyster from the Antwerp beds, which is figured by M. Nyst under the name *ungulata,* var. A.

M. Nyst says of this shell (p. 326 of his work), " La var. A est plus bombée.   Les sillons longitudinaux ont entièrement disparu sur les deux valves," but in his figure he has represented these " sillons " (radiations) obsolete or obscure, like they are on our Cor. Crag. shell.   He gives the localities of *O. ungulata* as Anvers and Bognor, but does not specify the special locality for var A.   The form in his pl. xxiv, fig. 1, is, however, probably *O. Bellovacina* from Bognor, while var. A is presumably from Anvers ; and on that assumption I have referred our Crag. shell to it, for it is certainly not the Eocene *Bellovacina.*

In the ever recurring difficulty as to whether shells in the Red Crag belong to that formation, or are only derivative in it, it is impossible to say whether this shell, of which specimens have occurred in the Red Crag, belongs to the age of that Crag or not ; but I have not met with the true form of the British *O. edulis* in the Red Crag.

I do not think now that the shell figured in my first Supplement, Tab. VIII, as *Ostrea plicatula* is the same as the shell here figured as *ungulata.*

MYTILUS EDULIS, var. GALLOPROVINCIALIS.   2nd Sup., Tab. VI, fig. 9.

MYTILUS GALLOPROVINCIALIS, *Lam.* An. Sans. Vert., t. vii, p. 46.
       —   —   *Phil.* Moll. Sic., vol. ii, p. 53, t. vi, figs. 12, 13.

Locality.   Red Crag, Sutton.

The specimen of this peculiar form, above figured, has been obtained by Mr. Edward Moore, of Woodbridge, from the Red Crag as above.

M<small>YTILUS</small> <small>EDULIS</small>, var. <small>UNGULATUS</small>.   2nd Sup., Tab. VI, fig. 9 *b.*

M<small>YTILUS</small> <small>UNGULATUS</small>, *Linn.* Syst. Nat., p. 1137.

Locality.   Cor. Crag? Boyton.

The present figure, *ungulatus*, represents a specimen obtained by Mr. Charlesworth, now in the cabinet of Dr. Reed; this is said to be from Boyton, and from the colour of the specimen, it most probably came from the Lower or Cor. Crag of that locality. These two very different forms of this genus, *galloprovincialis*, and *ungulatus*, are now generally admitted to be only variations of our common edible mussel, and I have introduced them to show that they lived in the Crag Sea. They were both figured by Dr. Jeffreys in the 'Mag. Nat. Hist.' for 1859, and at p. 10, *ungulatus* is there described as an "unquestionably distinct species;" but in his later work, the Brit. Conch., they are considered as varieties of *edulis*, in which opinion I coincide. Fig. 20, Tab. II, of 'Woodward's Geol. of Norfolk' is another form of this variable species.

P<small>ECTUNCULUS</small> <small>PILOSUS</small>, var. <small>INSUBRICUS</small>.   2nd Sup., Tab. VI, fig. 4 *a*, *b*; Crag. Moll., vol ii, Tab. IX, fig. 1 *d.*

A<small>RCA</small> <small>INSUBRICA</small>, *Broc.* Conch. Foss., sub. ap., p. 492, tav. xi, fig. 10 *a,b.*

Locality.   Cor. Crag, Sutton and Ramsholt.

When figuring the shells of this genus in 'Crag Mol., vol. ii, tab. ix, I gave a representation (fig. 1 *d*) of what I considered as an elongated variety of *P. glycimeris*, but this has since been given as a distinct species from the Crag, by Mr. A. Bell, as *P. insubricus*. I have therefore now given a figure of its interior, and I am unable to perceive any differences in this shell which justifies its separation from the general thick solid form which has been called *pilosus*, beyond its slightly more elongated form, and this may be connected with the more laterally extended form, common to *pilosus*, by individuals partaking more or less of this elongated character. The recent shell called *P. violacescens*, presents precisely the same form, with hinge and denticles the same. Fig. 5 *a* of Tab. IV is one of the laterally extended forms of *P. glycimeris*, from the Coralline Crag of Sutton, obtained by myself. Fig. 5 *b* is that of a specimen of my own from the Cor. Crag of Sutton, which seems to agree with that figured by Brocchi, 'Conch. Fos. Sub-Ap,' p. 483, Tab. II, fig. 8, under the name of *nummarius*. Fig. 4 *b* represents the inner lining of one of my specimens which separated itself; and as it corresponds with a figure given by Phillippi, 'En. Moll. Sic.,' Vol. II, Tab. XVIII, fig. 10 *a, b,* I thought it best to have it here figured.

Nucula turgens, *S. Wood*.   2nd Sup., Tab. V, fig. 6 *a, b*.

Spec. Char. *N. testá ovato-rotundatá, ventricosá, tumidá, partim lævigatá et partim con-centricè costulatá; margine dorsali et ventrali convexiusculá; margine intus denticulatá.*

Diameter ⅜ths of an inch.

Locality.   Red Crag, Waldringfield.

A single specimen of the genus *Nucula* is among Dr. Reed's specimens, kindly sent to me for examination, which I have here had represented; it has attached to it the name of *N. nucleus?* var.   I think, however, it cannot be referred to that species, which is much less inflated, and comparatively longer.   The two valves are closely united, and cannot be separated without endangering the integrity of the specimen.   The shell to which it seems to approach the nearest, from its tumidity, is *N. sphenoides*, Edwards, an Eocene species, but that shell differs in shape, being more angular and elongated. Our shell may be described as small, roundedly triangular, and very tumid, margin crenulated (the margins, though the valves are adherent, disclosing this).   The exterior, which has been much rubbed, is smooth on the part nearest the umbo, but deeply ridged on the part nearest the margin, and these ridges do not appear to be the result of decomposition.   Mr. Hancock has figured and described a shell under the name of *N. inflata*, 'Ann. and Mag. Nat. Hist.,' 1846, p. 333, pl. v, figs. 13, 14, and this, Mr. Hanley says, in his 'Monog. of the Nuculidæ' (p. 34, figs. 115, 116) is the same as *N. tenuis*, Möller (as he has determined from the examination of his specimen), but as this latter has a smooth margin and is more transverse than our present shell I am not able to refer the latter to it, and have therefore given to it provisionally a new name.

It may not improbably be a derivative specimen.

Arca tetragona, *Poli*.   2nd Sup., Tab. VI, fig. 8 *a, b*; Crag. Moll., vol. ii, p. 76, Tab. X, fig. 1; 1st Sup. to do., p. 116.

Locality.   Cor. Crag, Sutton.

The specimen 8 *b* now figured is given merely because it is that upon which the name of *Arca nodulosa*, Müll., was introduced by Mr. A. Bell, into his list of Crag shells in the 'Proc. of the Geological Association,' vol. ii.   It is now in the cabinet of Dr. Reed, and has been sent to me by that gentleman with the proposed name of *Arca puella*, A. Bell, attached. I have had a small specimen of my own finding here also represented (fig. 8 *a* of Tab. VI), which is very like it, and both, in my opinion, are specimens of *A. tetragona*, with coarser ornament than usual.

CHAMA GRYPHOIDES, *Linn., var.* GRYPHINA.   2nd Sup., Tab.V, fig.1 *a, b, c.*

CHAMA GRYPHOIDES, *Linn.*   Crag. Moll., vol. ii, p. 162, tab. xv, fig. 8.

Locality.   Red Crag, Waldringfield.

The specimen here represented is from Mr. Canham's collection. This I have referred as above, believing it to be merely a reversed form produced by the adherence of the right valve instead of the left. The present specimen is from the Red Crag, but probably only so by derivation from the Coralline.

LUCINA CRASSIDENS, *S. Wood.*   2nd Sup., Tab. V, fig. 4 *a, b.*

Diameter, ⅝ths of an inch.

Locality.   Red Crag, Waldringfield.

This is from Dr. Reed's cabinet; and it is in all probability a derivative from some anterior formation. The specimen seems to be not only full grown, but probably an old individual with a thickened interior. It has a prominent umbo, with a very broad and thickened hinge area. I thought at first sight that it might have been a specimen of *Lucina uncinata,* an Eocene species, which has an elevated dorsal margin, but that shell is much larger when full grown, and it has not the broad hinge of our shell. The present specimen is quite smooth on the exterior, but it has probably been much rolled and abraded.

Another specimen of this genus, from the nodule workings in the Red Crag, which, from having both valves adherent and filled with indurated material, is clearly also a derivative, was given to me by Mr. Charlesworth many years ago, and this I believe to be *Lucina crassa* from the Kimmeridge Clay.

LUCINOPSIS LAJONKAIRII, *Payr,* var. SUBOBLIQUA.   Figured in margin.

Locality.   Cor. Crag, Ramsholt.

A single valve of this species was found by myself some time ago, which in the outline differs so widely from all other specimens I have seen, that I have had it represented. It is suborbicular or slightly oblique, subequilateral, and much flatter

than the ordinary form : the exterior is covered with the same radiating fine striæ, decussated by lines of growth, as are present on the ordinary form, with which also its dentition is identical ; and it possesses the same impression or siphonal scar which is characteristic of *L. Lajonkairii.* As the differences presented by the present shell consist only in its greater flatness and different outline, I have regarded it as an accidental variety only ; but if a series should be obtained maintaining these characters, they might be regarded as of specific value, and the above name, *subobliqua*, then be assigned specifically.

Lucinopsis Lajonkairii, *Payr.* *var.* subobliqua, *S. Wood.*

ASTARTE MUTABILIS, *S. Wood.* 2nd Sup., Tab. VI, fig. 1.

ASTARTE MUTABILIS, *S. Wood.* Crag Moll., vol. ii, p. 179, tab. xvi, fig. 1.

Diameter, 2 inches.

Locality. Cor. Crag, near Orford.

I have had the present specimen figured for its great size, showing the margin without crenulations. This freedom from crenulation has always been considered by myself a distinguishing mark denoting that the animal which formed the shell had not arrived at maturity, and I can see no reason against such a supposition. This is as large as the largest of any specimens I have of this species, and larger than many which have the margin ornamented with crenulations. So far as I have studied the shells of the genus *Astarte*, I have always found the young or immature specimens of a species, that is decidedly crenulated when full grown, to be without that peculiarity.

In the plate of the "Arctic Shells," in Sir E. Belcher's 'Arctic Voyage,' are the figures of two species of *Astarte*. Fig. 7 *a, b,* of Tab. XXXIII, is named and described as new under the name of *A. Richardsoni.* This is stated by Dr. Jeffreys, in 'Ann. and Mag. Nat. Hist.' for 1877, p. 234, to be the same as *A. crebricostata* of Forbes, but unless the figure given in Belcher's work be erroneous, it seems to me to be the common form of *Astarte borealis*, such as occurs in the East Anglian beds ; while fig. 5 *a, b,* of the same Tab., called *A. fabula,* answers to the shell figured and described from the Red Crag as *A. crebrilirata,* 'Crag Moll.,' vol. ii, p. 184, tab. xvi, fig. 2, and which would thus appear to be living in the Arctic seas.

Mactra ponderosa ? *Stimpson.* 2nd Sup., Tab. VI, fig. 2.

Mactra ponderosa, *Stimpson.* Shells of New England.

Dimensions, 2 inches by 1¾.
Locality. Red Crag, Waldringfield.

A specimen of *Mactra* has been sent to me by Dr. Reed, with the above name and locality attached by (I believe) Mr. A. Bell. It is unknown to me either as recent or fossil, but it deserves a representation. Its form and appearance much resemble a large specimen of *M. solida*, and is different from *M. solidissima* (*M. ovalis*, Gould), 'Inv. Massach.,' p. 53, fig. 32, but it is not very far removed from it.

Mactra arcuata, *J. Sow.* Crag Moll., vol. ii, p. 243, Tab. XXIII, fig. 5; 1st Sup., p. 155.

I omitted to point out in my first Supplement that this species belongs to a section of the *Mactræ*, which the late Dr. J. E. Gray proposed to distinguish as a separate genus under the name of *Spisula*, this section being distinguished by the possession of the fimbriated mark or perpendicular striation on the lateral teeth, which forms part of the diagnosis of this species given at p. 243 of the 'Crag Moll.'; and that *Mactra glauca*, of which *arcuata* is called a variety in the list which accompanies Mr. Prestwich's paper on the Crag, belongs to the other section, viz. that which is destitute of this impression.

A fragment of a full-grown shell, showing the hinge with this fimbriated mark, and which therefore seems to be one of *M. arcuata*, was obtained by my son from a band of shell fragments at the top of the Middle Glacial sand, three or four feet below the over-lying chalky clay, in a well at Bealings, near Woodbridge, this seam exactly corresponding in position to that at Billockby and Hopton, from which the species given in my first Supplement were obtained.

Thracia papyracea, *Poli.* 2nd Sup., Tab. VI, fig. 6 *a, b*.

Thracia phaseolina. Crag Moll., vol. ii, p. 259, tab. xxvi, fig. 2.
— papyracea. 1st Sup. to do., p. 156.

Additional localities. Chillesford Bed, Sudbourn Church Walks; Lower Glacial, Belaugh.

Dr. Reed having sent to me a specimen with the name *Thracia villosiuscula* attached, upon which that name as a variety of *T. papyracea* had been introduced into the list by Mr. A. Bell in the 2nd vol. of the ' Proceedings of the Geol. Association,' I have had it figured as above (6 *b*), and with it one of my own from the same locality, exhibiting the ordinary form of *papyracea* (6 *a*).

T. villosiuscula is considered both by Forbes and Hanley and by Dr. Jeffreys as a variety of *papyracea*, as being more equilateral than the typical form of that species, but the specimen sent me by Dr. Reed is rather less equilateral than the typical form. The species itself is difficult of distinction from the young of *T. pubescens*.

I also possess a perfect specimen of this shell from the Lower Glacial sand of Belaugh.

THRACIA VENTRICOSA, *Phil.* 2nd Sup., Tab. V, fig. 3 ; Crag Moll., vol. ii, p. 262, Tab. XXVI, fig. 5 ; 1st Sup., p. 156.

Locality. Cor. Crag, Ramsholt.

In the list of the Crag shells appended to Mr. Prestwich's paper, p. 141, the one I called by the above name is said to be *Thracia convexa*, W. Wood, and I have in consequence figured a specimen obtained by myself from the Cor. Crag of Ramsholt. I thought, and still think, that the differences between the Crag shell and *T. convexa* are sufficient for their being kept distinct, and the specimen now figured exhibits these differences better than that figured in my original work ; they consist in *ventricosa* having a far greater length of the posterior part of the shell and a less tumidity of the anterior. Indeed, the form of *ventricosa* is nearer to that of *pubescens* than it is to *convexa*.

In this, as in many other similar cases of living species approaching the Crag form, *T. convexa* may be the descendant of *T. ventricosa*, but if so the time which has elapsed since the deposit of the Coralline Crag has been sufficient to produce those differences, which, as I have pointed out in the concluding remarks of my first Supplement (p. 192), I consider should justify us in designating species as distinct.

PHOLAS INTERMEDIA, *S. Wood.* 2nd Sup., Tab. VI, fig. 7 ; Tab. V, fig. 2 *a—c.*

Dimensions. Length, 2 inches. Breadth of valve, $1\frac{3}{16}$ inch.
Localities. Cor. Crag, Gedgrave ; Red Crag, Waldringfield.

The specimen represented in Tab. V, fig. 2, is in the collection of Mr. Canham, now in the Ipswich Museum, and was obtained from the phosphatic nodule pits at Waldringfield. As the valves are held together by the Red Crag material within them, I infer that the specimen died in the Red Crag, the material of which occupied the cavity as the animal decayed, though the valves are not precisely adherent as they are in life. But for this I should have supposed it to have been a derivative from the Coralline Crag, from which the smaller specimen shown in fig. 7 of Tab. VI was obtained. I at first thought that it might be the same as the *Pholas brevis* from the Cor. Crag, of which I was enabled to figure a fragment in my first Sup. (Tab. X, fig. 24); but the differences are so great that I cannot regard the two as identical. Both shells, however, belong to the true genus *Pholas,* and not to that section of it called *Zirphea,* which was proposed as a separate genus by the late Dr. J. E. Gray; and in which the rays are confined to the anterior portion of the shell, and are bounded by a deep sulcus; and to which section *P. crispata* belongs.

The specimen, consisting of a single and smaller valve, which is represented in Tab. VI, fig. 7, was sent to me by Dr. Reed, with the name of *Pholas parva* attached, as from the Coralline Crag of Gedgrave, but it seems so closely to resemble the large shell from the Red Crag, represented in Tab. V, fig. 2, that I think it must be the younger state of it. It differs from *parva* in being considerably shorter in proportion to its breadth, the figure of that species from the Red Crag, given in the first Sup., Tab. X, being taken from a specimen which had been somewhat distorted by confinement in the crypt, and I have not seen that species in the Coralline Crag. I think it possible that the small specimen represented in fig. 24*b* of Tab. X of my first Supplement, may be a still younger state of our present shell instead of, as supposed in that Supplement, the young of the shell represented in fig. 24*a* of the same plate (and which I retain as *Pholas brevis*), as it has a similar deep opening for the foot; but a good series is required for a satisfactory determination of that question.

Pholas dactylus, Linn., has been given as a species from the Red Crag in Mr. Prestwich's paper 'Quart. Jour. Geol. Soc.,' vol. xxvii, p. 485, and by Mr. Bell in his paper on the English Crags, 'Proc. Geol. Assoc.,' vol. ii, No. 5, p. 26, from the "Middle (or Oldest Red) Crag." I have procured from Dr. Reed the specimen upon which this identification was based, and which has the locality of Walton Naze marked upon it, and to set the subject at rest I have had it represented in fig. 5 of Tab. V. The specimen exhibits unequivocally those characteristics which I have pointed out at p. 295 of the second volume of the 'Crag Mollusca' as distinguishing *cylindrica* from *dactylus,* and there can be no question of its being the common Walton species, *Ph. cylindrica,* J. Sow. In the list[1] given in the lately published memoir of the 'Geol. Survey,' for half sheet No. 48, this species is introduced, but this is probably only by adoption from the

[1] There are some errors in this list, even as regards Walton; but that part of it which refers to Beaumont (and which I presume is merely a repetition of the late Mr. John Brown's list of shells obtained from that locality) is, in my opinion, quite untrustworthy. *Pyrula uniplicata,* Duj., given in the memoir list, is probably a clerical error for some other shell, possibly *Pyramidella unisulcata,* which in Mr. Prestwich's *Coralline* Crag list is regarded as identical with *P. læviuscula,* but which I do not consider to exist in any part of the Crag. There is also a clerical error in respect of that shell in Mr. Prestwich's *Red* Crag list.

list in Mr. Prestwich's paper, in which the name was introduced from the specimen of *cylindrica* now under consideration. *Pholas lata* is also given in the same memoir as from Beaumont, but I do not know such a species unless it be *Pholas crispata*, to which shell the name of *lata* was given by Lister (see the synonyms of that shell in vol. ii of 'Crag. Moll.,' p. 296).

Venus dysera, Brocchi, and *Venus fasciata*, Dacosta, are given by Mr. A. Bell from the Cor. Crag, but I believe the former of these to be the young state of *Venus imbricata*, a specimen of which I had represented in 'Crag Moll.,' vol. ii, Tab. XIX, fig. 3 *b*. This may *possibly* be, in the young condition, undistinguishable from *V. fasciata*, but I have not yet seen any specimen from the Cor. Crag that could be pronounced positively as identical with that species. The young of many proximate but distinct species so closely resemble each other as to be incapable, in that state, of separation, the specific distinction only appearing as the animal advances in growth. I cannot therefore admit *dysera* into my list at all, nor *fasciata* into it as a Cor. Crag shell.

BRACHIOPODA.

DR. JEFFREYS has recently described several species of *Brachiopoda* that were obtained by the deep-sea dredgings during the expeditions of H.M.S. "Lightning" and "Porcupine," and he has figured them in the 'Proceedings of the Zool. Soc.,' April 16th, 1878. One of these species, to which he has given the name of *Terebratula trigona*, Plate xxii, fig. 3, very strongly resembles a small specimen that I found in the Cor. Crag of Sutton, and which is figured in my first Supplement, Tab. xi, fig. 3 *c*, and there considered as a young or small variety of *Terebratulina caput serpentis*, and I am disposed to think that if the crag fossil could be compared with the recent shell they might perhaps be specifically united. I cannot say if there be any difference in the form of the loop in my specimen, as I am unable to separate the valves of the only one at present known to me. I have also figured another specimen from the Cor. Crag in the same plate (fig. 3 *d*) as *caput serpentis*, but this is so abnormal that when more and similar specimens are found it may be perhaps entitled to specific distinction, and be called *anceps*. At p. 169 of my first Supplement I have pointed out that the beak of this latter shell has the form of that possessed by *Rhynconella*. In the 'Quarterly Journ. of the Geol. Society,' vol. xxvii, p. 137, Dr. Jeffreys says that the *Discina* from the Cor. Crag is the same species as *Discina Atlantica*, King; possibly this may be so, but, as in the case of the above *Terebratulina*, better evidence than we at present possess will be necessary for the correct determination of the question. The only two specimens of the Crag *Discina* that I know, or have heard of, were found by myself, and these are both upper valves. One of them is that figured by Mr. Davidson in 1852, also in Tab. XI of my first Supplement, and is in the collection of Crag Mollusca which I gave to the British Museum, and this is not perfect. The other (which is in my own cabinet) I found subsequently, and in this the characters are obscured by the shell being covered with a mass of *Cellepora*.

MEMORANDUM.

THE following species, all contained in my original synoptical list, have also since occurred in beds represented in its columns beyond what is there shown.

IN THE RED CRAG OF SUTTON AND BUTLEY.—*Nassa conglobata*. A solitary specimen found at Walton thirty-five years ago by Mr. Charlesworth, and in my collection in the British Museum, was the only instance of this shell known to me until lately. In Mr. Canham's collection, however, I observed a specimen from the Red Crag of Sutton; and it seems to me, therefore, that although it has not yet occurred in the Coralline Crag, this shell is properly a species of that Crag, and not of the Red, and is only present in the latter (albeit that it has occurred at Walton) by derivation from the Coralline.

IN THE CHILLESFORD BEDS.—*Cardita corbis* and *Abra prismatica*. Mr. Dowson informs me that he has found several specimens of these shells at Aldeby.

IN THE LOWER GLACIAL.—From a fossiliferous seam in the pebbly sands near Southwold Mr. Crowfoot has obtained several of the species given in my original list from these sands in Norfolk, and in addition *Cerithium tricinctum*, *Melampus* (Conovulus) *pyramidalis*, and *Donax vittatus*. Perfect specimens also of the latter from Belaugh and Weybourn are in my cabinet. An imperfect specimen of *Cardium* in my cabinet from Belaugh seems referable to *Cardium Islandicum*, but no reliance can be placed upon such fragments, either in this or other beds, for specific determination. Similarly, the fragments upon which the name of *C. Grœnlandicum* is inserted in the list of shells given by Mr. C. Reid from these sands where they underlie the Till along the Cromer coast (in the 'Geological Magazine' for July, 1877), are equally unreliable, and might be referred to more than one large species of *Cardium*. Whether *Islandicum* or *Grœnlandicum*, the Belaugh and Cromer fragments are probably those of the same species only, and would answer as well for the one as for the other of these shells. Mr. Crowfoot also gives the name *Grœnlandicum* among those of the species obtained by him from the pebbly sands at Southwold. *Astarte sulcata, Ostrea edulis,* and *Pleurotoma turricula* are also given by Mr. C. Reid as having been found by him in these sands on the Cromer coast.

In the Middle Glacial.—*Hydrobia ulvœ.* A specimen of this shell was found by Mr. Harmer at Lound, near Yarmouth, in association with some of the commoner species of this deposit.

In the March Gravel.—*Tellina lata.* A small specimen of this shell from March is in the Cambridge Museum. Mr. Harmer has found the freshwater shell, *Cyrena fluminalis,* in numbers in this gravel, associated with *Cardium edulis* and other marine shells; an asociation corresponding to that which occurs in the Hessle gravel at Kelsea Hill in Yorkshire.

ADDITION TO THE SYNOPTICAL LIST GIVEN AT PAGE 203 OF FIRST SUPPLEMENT TO "THE CRAG MOLLUSCA."

Species and varieties new to the Synoptical List are in Roman letters. Species already in the Synoptical List are in italics, and are only inserted to indicate their occurrence in some one or other of the formations, referred to in the separate columns, beyond what is specified in the original list. Such of the latter as are marked † are given in the Lower Glacial column, on the authority *only* of Mr. C. Reid's paper, on the "Cromer Pliocene," in the 'Geological Magazine' for July, 1877.

Page in 2nd Supp.		Cor. Crag.	Red Crag, Walton.	Red Crag, Sutton and Butley.	Scrobicularia Crag.	Fluvio-marine Crag.	Chillesford Beds.	Lower Glacial.	Middle Glacial.	Upper Glacial.	Post Glacial, Kelsey.	Post Glacial, March.	Post Glacial, Hunstanton.	Post Glacial, Nar Brickearth.	Living, Britain.	Living, Mediterranean.	Living, elsewhere.	REMARKS.
4	Columbella sulculata, *S. Wood*	...	...	×	...	...	...	...	...	...	...	...	...	...	...	...	...	
2	*Nassa prismatica, Broc.*	×	...	...	...	...	...	...	...	...	...	...	...	...	...	×	×	
4	— incrassata, Müll., var. tumida.	...	...	×	...	...	...	...	...	...	...	...	...	...	...	...	...	
4	— angulata? *Broc.*	×	...	...	...	...	...	...	...	...	...	...	...	...	...	...	...	
52	— conglobata, Broc.	?	...	×	...	...	...	...	...	...	...	...	...	...	...	...	...	Derivative in Red Crag of Sutton; possibly derivative also at Walton.
3	— microstoma, S. Wood (*N. prismatica*, var. limata).	×	×	×	...	...	...	...	...	...	...	...	...	...	...	...	...	
2	*Buccinum Dalei? J. Sow.*, var. distorta.	×	...	...	...	...	...	...	...	...	...	...	...	...	...	...	...	
1	— *undatum*, Linn., var. distorta	...	...	×	...	...	...	...	...	...	...	...	...	...	...	...	...	
	— — var. *tenerum*	...	...	...	...	...	...	×	...	...	...	...	...	...	...	...	...	B. Grœnlandicum, Chem. sec. Jeffreys Ann. and Mag. Nat. Hist., 1876, p. 323.
1	— nudum, S. Wood	×	...	..	...	...	...	...	...	...	...	...	...	...	...	...	...	
2	— declive, S. Wood	×	...	...	...	...	...	...	...	...	...	...	...	...	...	...	...	
13	Murex Reedii, *S. Wood*	×	...	...	...	...	...	...	...	...	...	...	...	...	...	...	...	
15	— recticanalis, *S. Wood*	×	...	...	...	...	...	...	...	...	...	...	...	...	...	...	...	
15	— Crowfootii, *S. Wood*	×	...	...	...	...	...	...	...	...	...	...	...	...	...	...	...	
14	— pseudo Nystii, *S. Wood*	×	...	...	...	...	...	...	...	...	...	...	...	...	...	...	...	
16	Ranella? Anglica, *A. Bell*	...	...	×	...	...	...	...	...	...	...	...	...	...	...	...	...	Derivative.
15	*Triton connectens?* S. Wood	...	...	×	...	...	...	...	...	...	...	...	...	...	...	...	...	Derivative in Red Crag.
9	Fusus Waelii, *Nyst*	×	...	...	...	...	...	...	...	...	...	...	...	...	...	...	...	
11	— obscurus, *S. Wood*	×	...	...	...	...	...	...	...	...	...	...	...	...	...	...	...	
12	— nodifer, *A. Bell*	...	...	×	...	...	...	...	...	...	...	...	...	...	...	...	...	
11	— ? exactus, *S. Wood*	×	...	...	...	...	...	...	...	...	...	...	...	...	...	...	...	Derivative.
6	Trophon(Sipho)Islandicus, *Gmel.*	...	...	×	...	...	...	...	...	...	...	...	...	×	...	...	×	
7	— (—) gracilis, Da Costa	×	...	×	...	...	...	...	...	...	...	...	...	...	...	...	×	
6	— (—) tortuosus, L. Reeve	...	...	×	...	...	...	...	...	...	...	...	...	...	...	...	×	
7	— (—) Olavii, *Beck*	...	...	×	...	...	...	...	...	...	...	...	...	...	...	...	×	
9	— Kröyeri, *Moll.*	...	...	×	...	...	...	...	...	...	...	...	...	...	...	...	×	
8	— pseudo Turtoni, *S. Wood*	...	...	×	...	...	...	...	...	...	...	...	...	...	...	...	...	Trophon Norvegicus of the synoptical list.
16	Pleurotoma Morreni, *De Koninck*	...	...	×	...	...	...	...	...	...	...	...	...	...	...	...	...	Derivative at Waldringfield.
18	— *teres*, Forbes	×	...	...	...	...	...	...	...	...	...	×	...	...	...	...	...	
18	— gracilicostata, *S. Wood*	×	...	...	...	...	...	...	...	...	...	...	...	...	...	...	...	
17	— curtistoma, *A. Bell*	×	...	...	...	...	...	...	...	...	...	...	...	...	...	...	...	
21	— pannus, *Bast.*	×	...	...	...	...	...	...	...	...	...	...	...	...	...	...	...	
20	— *senilis*, S. Wood	...	...	×	...	...	...	...	...	...	...	...	...	...	...	...	...	
20	— catenata, *A. Bell*	×	...	...	...	...	...	...	...	...	...	...	...	...	...	...	...	Derivative.
†	— *turricula*, Mont.	...	...	...	...	...	...	×	...	...	...	...	...	...	...	...	...	Possibly the same as *pannus*.
22	Cancellaria (Admete) Avara? *Say*	...	...	×	...	...	...	...	...	...	...	...	...	...	...	...	...	
22	— crassistriata, *A. Bell*	...	...	×	...	...	...	...	...	...	...	...	...	...	...	...	×	
39	Cerithium derivatum, *S. Wood*	...	...	...	...	×	...	...	...	...	...	...	...	...	...	...	...	Derivative.
23	— Greenii? *Adams*	...	...	...	...	×	...	...	...	...	...	...	...	...	...	...	×	Derivative.
52	— *tricinetum.* Broc.	...	...	...	...	...	×	...	...	...	...	...	...	...	...	...	...	
27	Turritella Taurinensis, *Mich.*	...	...	×	...	...	...	...	...	...	...	...	...	...	...	...	...	Derivative.
27	— incrassata, var. acutangula? Broc.	×	...	...	...	...	...	...	...	...	...	...	...	...	...	...	...	
27	— — var. subangulata, Broc.	×	...	...	...	...	...	...	...	...	...	...	...	...	...	...	...	
25	Scalaria torulosa, *Broc.*	×	...	...	...	...	...	...	...	...	...	...	...	...	...	...	...	
26	— geniculata? *Broc.*	×	...	...	...	...	...	...	...	...	...	...	...	...	...	...	...	

Page in 2nd Supp.		Cor. Crag.	Red Crag, Walton.	Red Crag, Sutton and Butley.	Scrobicularia Crag.	Fluvio-marine Crag.	Chillesford Beds.	Lower Glacial.	Middle Glacial.	Upper Glacial.	Post Glacial, Kelsey.	Post Glacial, March.	Post Glacial, Hunstanton.	Post Glacial, Nar Brickearth.	Living, Britain.	Living, Mediterranean.	Living elsewhere.	REMARKS.
24	Chemnitzia senistriata, *S. Wood*	×	...	...	...	...	...	...	...	...	...	...	...	...	...	...	...	
24	— *internodula*, S. Wood, var. ligata.	...	...	...	...	×	...	...	...	...	...	...	...	...	...	...	...	
39	Odostomia derivata, *S. Wood*	...	...	...	...	×	...	...	...	...	...	...	...	...	...	...	...	Derivative.
28	Eulima Hebe, *Semper*	×	...	...	...	...	...	...	...	...	...	...	...	...	...	...	...	
29	— *subula*, D'Orb	×	...	...	...	...	...	...	...	...	...	...	...	...	...	...	...	
27	— Naumanni ? *von Könen*	×	...	...	...	...	...	...	...	...	...	...	...	...	...	...	...	
28	— robusta, *A. Bell*	...	...	×	...	...	...	...	...	...	...	...	...	...	...	...	...	Derivative ?
29	Rissoa parva ? *Da Costa*	×	...	...	...	...	...	...	...	...	...	...	...	...	×	×	×	
29	— *costulata*, Alder	×	...	...	...	...	...	...	...	...	...	...	...	...	×	×	×	
40	— *proxima*, Alder	...	...	...	...	×	...	...	...	...	...	...	...	...	...	...	...	
35	Assiminea Grayana, *Leach*	...	...	...	...	×	...	...	...	...	...	...	...	...	×	...	...	
30	Hydrobia obtusa, *Sandberger*	...	...	...	...	×	...	...	...	...	...	...	...	...	...	...	...	
53	— *ulvæ*, Penn	...	...	...	...	...	...	...	×	...	...	...	...	...	...	...	...	In the middle glacial of Lound.
33	Amaura hesterna, *S. Wood*	?	...	?	...	...	...	...	...	...	...	...	...	...	...	...	...	{ Doubtful whether from the Cor. Crag or from the Red of Butley, &c.
30	Natica (Amauropsis) Japonica ? *A. Adams.*	...	...	×	...	...	...	...	...	...	...	...	...	...	...	...	×	
31	— triseriata ? *Say*	...	...	×	...	...	...	...	...	...	...	...	...	...	...	...	×	
32	— *helicina* var heliciformis, S. Wood	×	...	...	...	...	...	...	...	...	...	...	...	...	...	...	...	
31	— *Groenlandica*, Beck, var. declivis.	...	...	×	...	...	...	...	...	...	...	...	...	...	...	...	...	
34	Adeorbis ? naticoides, *S. Wood*	×	...	...	...	...	...	...	...	...	...	...	...	...	...	...	...	A very doubtful species.
38	Melampus fusiformis, *S. Wood*, var. elongatus.	...	...	×	...	...	...	...	...	...	...	...	...	...	...	...	...	Derivative ?
52	— *pyramidalis*, J. Sow.	...	...	...	...	...	×	...	...	...	...	...	...	...	...	...	...	{ Rackheath; Southwold; and, according to Mr. Reid, Runton.

BIVALVIA.

Page in 2nd Supp.		Cor. Crag.	Red Crag, Walton.	Red Crag, Sutton and Butley.	Scrobicularia Crag.	Fluvio-marine Crag.	Chillesford Beds.	Lower Glacial.	Middle Glacial.	Upper Glacial.	Post Glacial, Kelsey.	Post Glacial, March.	Post Glacial, Hunstanton.	Post Glacial, Nar Brickearth.	Living, Britain.	Living, Mediterranean.	Living elsewhere.	REMARKS.
41	Ostrea ungulata, *Nyst*	×	×	×	...	...	...	...	...	...	...	...	...	...	...	...	...	
53	† — *edulis*, Linn.	...	...	...	...	...	...	×	...	...	...	...	...	...	...	...	...	
43	*Mytilus edulis, var.* ungulatus	×	...	...	...	...	...	...	...	...	...	...	...	...	...	...	...	
42	— — *var.* galloprovincialis	...	...	×	...	...	...	...	...	...	...	...	...	...	...	...	...	
43	*Pectunculus pilosus, var.* insubricus, Broc.	×	...	...	...	...	...	...	...	...	...	...	...	...	...	...	...	
43	— *var.* nummarius, Broc.	×	...	...	...	...	...	...	...	...	...	...	...	...	...	...	...	
44	Nucula turgens, *S. Wood*	...	...	×	...	...	...	...	...	...	...	...	...	...	...	...	...	Derivative ?
45	Lucina crassidens, *S. Wood*	...	...	×	...	...	...	...	...	...	...	...	...	...	...	...	...	Derivative.
45	*Lucinopsis Lajonkairii*, Payr., var. subobliqua.	×	...	...	...	...	...	...	...	...	...	...	...	...	...	...	...	
45	*Chama gryphoides, Linn.*, var. gryphina.	...	...	×	...	...	...	...	...	...	...	...	...	...	...	...	...	Derivative in Red Crag.
52	*Cardita cordis*, Phil.	...	...	...	...	...	×	...	...	...	...	...	...	...	...	...	...	Aldeby.
52	*Cardium Islandicum* ? Linn.	...	...	...	...	...	...	×	...	...	...	...	...	...	...	...	...	
53	†*Astarte sulcata*, Da Costa	...	...	...	...	...	...	×	...	...	...	...	...	...	...	...	...	
52	*Donax vittatus*, Da Costa (anatinus. F. & H.)	...	...	...	...	...	...	×	...	...	...	...	...	...	...	...	...	{ Belaugh and Weybourne; also, according to Mr. C. Reid, from Runton.
53	*Tellina lata*, Gmel.	...	...	...	...	...	...	...	...	...	...	×	...	...	...	...	...	{ A specimen from the March gravel in the Cambridge Museum.
52	*Abra prismatica*, Mont.	...	...	...	...	...	×	...	...	...	...	...	...	...	...	...	...	Several specimens from Aldeby.
47	Mactra ponderosa ? *Stimpson*	...	...	×	...	...	...	...	...	...	...	...	...	...	...	...	...	
47	— arcuata, J. Sow.	...	...	...	...	...	...	...	×	...	...	...	...	...	...	...	...	
47	*Thracia papyracea*, Poli	...	...	...	...	...	...	×	...	...	...	...	...	...	...	...	...	Perfect from Belaugh.
48	Pholas intermedia, *S. Wood*	×	...	×	...	...	...	...	...	...	...	...	...	...	...	...	...	

The following species should be omitted from the Synoptical List altogether, viz. *Trophon Norvegicus*, see p. 7; *Pleurotoma violacea*, see p. 20; and *Pholas dactylus*, see p. 49; and from the Coralline and Red Crag columns of the list, *Ostrea edulis*, see p. 42.

INDEX

SECOND SUPPLEMENT OF THE CRAG MOLLUSCA.

PLATE I.

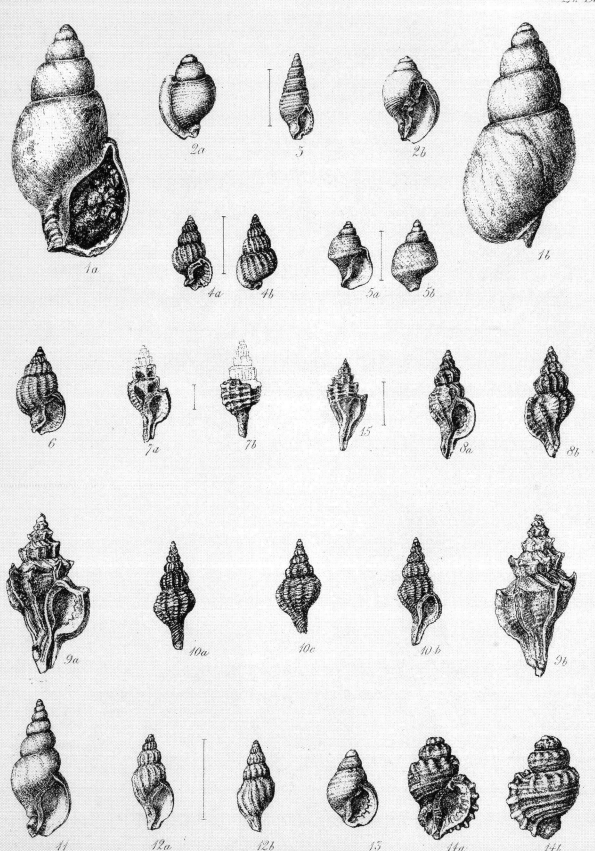

PLATE II.

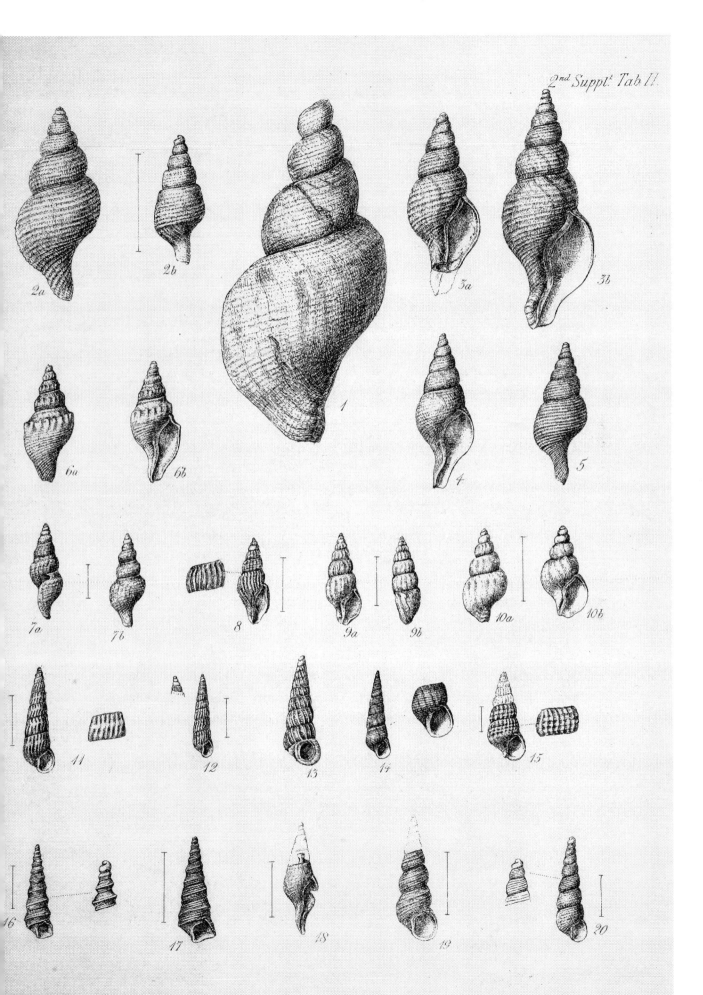

PLATE III.

* Referred to at p. 9 as Tab. III, fig. 8.
† Referred to at p. 5 as Tab. III, fig. 11.

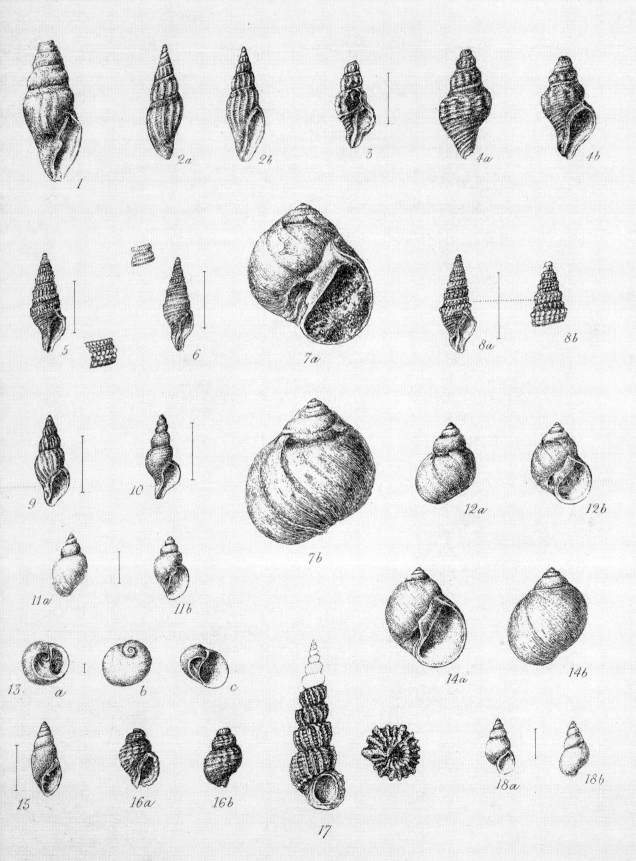

PLATE IV.

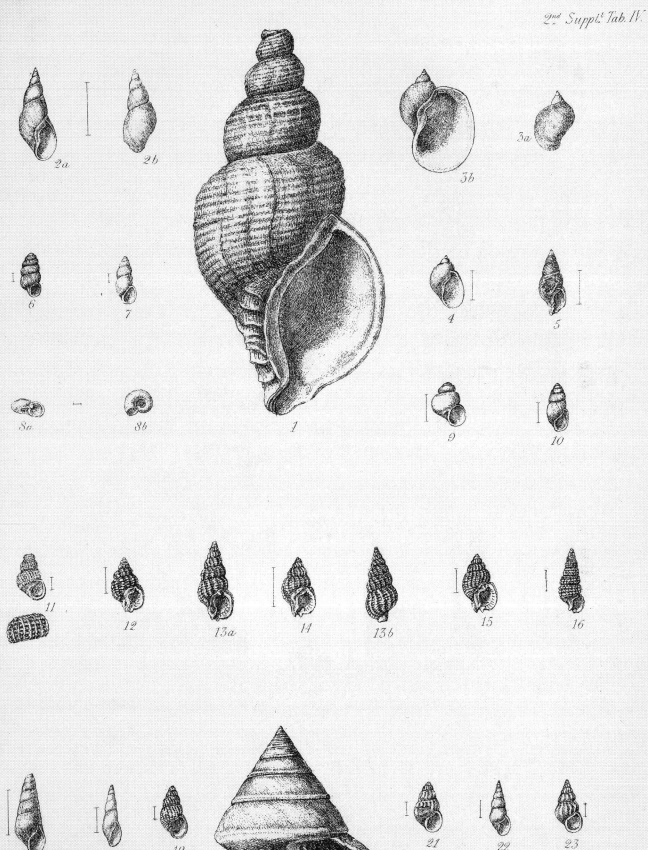

2ⁿ Suppᵗ Tab.IV.

G. B. Sowerby.

PLATE V.

The inside view of this oyster, not having been reversed by the Engraver, fig. 7 *b* presents an erroneous appearance, inasmuch as that the umbo of the valve should turn to the right instead of the left. Viewed by reflection in a mirror the representation will be found correct.

1a
1b
1c
2a
2b
4a
4b
2c
3
6a
6b
5
7a
7b

G. B. Sowerby

PLATE VI.

1

2

3a

3d

3e

4b

4a

5a

3b

3c

6b

6o

5b

3f

7

8a

8b

9a

9b

G. B. Sowerby

THE

PALÆONTOGRAPHICAL SOCIETY.

INSTITUTED MDCCCXLVII.

VOLUME FOR 1882.

LONDON:

MDCCCLXXXII.

THIRD SUPPLEMENT

TO THE

CRAG MOLLUSCA,

COMPRISING

TESTACEA FROM THE UPPER TERTIARIES OF THE EAST OF ENGLAND.

BY THE LATE

SEARLES V. WOOD, F.G.S.

EDITED BY HIS SON SEARLES V. WOOD, F.G.S.

PREFACE; PAGES 1—24; PLATE I.

UNIVALVES AND BIVALVES.

LONDON:

PRINTED FOR THE PALÆONTOGRAPHICAL SOCIETY.

1882.

PRINTED BY
J. E. ADLARD, BARTHOLOMEW CLOSE.

PREFACE BY THE EDITOR.

My late father had at the time of his death (which took place on Oct. 26th, 1880) collected some materials and written the text for a further short Supplement to his original work on the "Crag Mollusca." These materials and text consisted of the descriptions here given, and also of those of the remains of certain vermiform mollusca which he had got together from the Coralline and Red Crag beds. The latter, however, were not left by him in such a form as would allow me to give his views without risk of misrepresentation; and as I know, moreover, that in respect of one at least of these forms he was in great doubt to the last, whether it belonged to the Molluscous sub-kingdom at all, I have thought it best to suppress that portion of his notes, and to give only the portion which relates to the Gasteropoda and Bivalvia; as to which I well know what his ideas were. This part forms but an insignificant addition to the preceding portions of his work, and comprises for the most part only shells that have got into the Red Crag beds by derivation from older formations; but as all such shells must be considered, and eliminated from the evidence which is obtainable to show of what the molluscan Fauna of that part of the North Sea which washed the shore of East Anglia at the time of the Red Crag really consisted, their description and representation by figure, as my father intended, appear to me to form a proper sequel to his work.

The text which is not comprised by brackets is that left by my father. The text within brackets (with the exception of the description of *Margarita crassi-striata*, and of the bed at Boyton from which that shell was obtained, which is by Mr. Robert Bell, is by myself.

THIRD SUPPLEMENT TO THE CRAG MOLLUSCA.

GASTEROPODA.

Rostellaria? gracilenta, *S. Wood*. 3rd Sup., Tab. I, fig. 1.

Axis, 1 inch.

Locality. Red Crag, Felixstowe.

Many years ago I found a few specimens in the Red Crag at Sutton, to which I gave the provisional name of *Rostellaria plurimacosta* in my original Catalogue in ' Mag. Nat. Hist.,' September, 1842, p. 543. Not finding any of the like form and character in better preservation I, in the first supplement (1872) to my work on the ' Crag Mollusca ' (p. 5, Tab. II, fig. 14), gave a figure with the best information I possessed respecting the few specimens in my own cabinet, and referred them (doubtfully) to a well-known Eocéne species *R. lucida*, J. Sow.

In my recent researches at Felixstowe I have obtained three or four more specimens of this shell, though in a more mutilated condition. With these I have found some other mutilated specimens, the best of which I have here had figured. This resembles in its ornamentation the Eocene species *lucida*, which is from the upper part of the London Clay (' Min. Con.,' Tab. 91), but it differs in other respects, as it is much more slender, more elongated, and possesses larger and fewer costulæ. Unfortunately the mouth or aperture is imperfect so that the genus cannot with certainty be determined. I, however, propose for it provisionally the name above. It is undoubtedly an immature specimen, with its outer lip sharp as it would naturally be in a young and growing shell.

In the Ipswich Museum there is a mass of material, nearly two feet across and about three inches in thickness, found in the nodule bed at the base of the Red Crag at Waldringfield, and on the upper surface are a large number of specimens of a vermiform shell identical with what has been figured in ' Min. Conch.,' Tab. 596, figs. 1—3, as *Vermetus Bognoriensis*, and with them are several specimens, but in a mutilated condition, of what may be referred to *Rostellaria lucida*, as also some specimens resembling my present shell in a similar condition to my own above figured. There can

1

therefore, I think, be little doubt but that the shell now figured is like the true *lucida*, a London Clay species, and has got into the Crag by derivation from that formation ; for the shell figured by Sowerby in Dixon's ' Geology of Sussex,' Tab. V, fig. 21, from the Bracklesham beds as *R. lucida*, differs from that originally figured by him under this name in ' Min. Con.' (and which was from the London Clay of Highgate), and, in my opinion, is specifically distinct from it, as it possesses more numerous and sharp ribs or costulæ, and is more regularly striated in a spiral direction, the striations covering the entire surface.

TROPHON ANTIQUUS, var. DESPECTUS.   3rd Sup., Tab. I, fig. 9.

MUREX DESPECTUS, *Linn.*   Syst. Nat., edit. xii, p. 1222, 1766.
FUSUS   —   *Lam.*   An. sans Vert., 2nd ed., tom. ix, p. 448, 1843.
  —   —   *Fleming.*   Brit. Anim., p. 349, 1828.
TRITONIUM DESPECTUM var. ANTIQUATA, *Middendorf.*   Malkop., p. 135, 1849.

Locality.   Red Crag, Sutton.

In the first portion of my work I have given many of the extreme forms of this variable species, but there is no figure representing the front or opening of the present variety ; and as the above name of *despectus* has been several times given as a distinct species from the Red Crag I have thought it necessary to represent a shell here which resembles the recent form of that name.   This was introduced as a distinct Crag species by the late Sir Chas. Lyell in a list accompanying a paper by him, and published in the ' Mag. Nat. Hist.' in 1839, p. 329 ; by the late Edward Forbes, also, in his Memoir in the ' Geol. Survey,' 1846, p. 426, and in the list by Professor Prestwich in ' Quart. Journ. Geol. Soc.,' vol. xxvii, p. 488.   I think it therefore incumbent on me to give the accompanying figure of this variety, for such only do I conceive it to be.   I will, therefore, refer to Plate V of my first volume, and assign the figures therein as the following varieties of this species according to my view, viz. *Fusus decemcostatus*, Gould, ' Invert. Massach.,' is represented in it by fig. 1 *a ; Fusus carinatus*, Lam., by fig. 1 *b ; Fusus striatus*, Sow., by fig. 1 *c ; Fusus contrarius*, Phil. and Nyst, by figs. 1 *d—k*.

There are some other varieties, I believe, in the Crag of which I have not been able to obtain specimens for representation.   *Fusus tornatus*, Gould, is, I believe, only a variety of *T. antiquus*, and the shell figured in the ' Ency. Method.' with wavy ridges, pl. 426, fig. 4, is another variety, and this I am told has been found in the Red Crag, but I have not been able to see a specimen or I would have had it figured.   Brown, in his ' Illustr. Brit. Conch.,' pl. 47, figs. 10 and 13, has figured this shell with wavy ridges, and calls it *Fusus subantiquatus*, but says, " I have great doubts of this being a British shell." This undulation is produced by a sinuated form of the outer lip, and is probably a distortion, and if so the specimens are not likely to be very numerous.

TROPHON MURICATUS, *Mont.* Crag Moll., vol. i, p. 50, and 1st Sup., p. 28.

TROPHON MURICATUS var. EXOSSUS. 3rd Supplement, tab. i, fig. 3, 1882.

Locality. Red Crag, Felixstowe.

The specimen figured as above was recently found by me, and though in excellent preservation is quite destitute of the longitudinal ribs present in the ordinary form of this species. I have therefore distinguished it as a variety, under the name of *exossus.*

PLEUROTOMA TURRIS, *Lamarck*, 3rd Sup., Tab. I, fig. 8.

PLEUROTOMA TURRIS, *Lam.* An. sans Vert., tom. vii, p. 97, 1822.
— — — Ibid., 2nd ed., tom. ix, p. 367, 1843.
— — — Ency. Method., p. 795, t. 441, fig. 7, 1832.
— — *Nyst.* Coq. foss. de Belg., p. 525, 1843.
MUREX INTERRUPTUS, *Brocchi.* Conch. foss. Subap., p. 433, pl. ix, fig. 21, 1814(

Spec. Char. "*T. fusiformi-turrita, transversim sulcato-rugosa; striis longitudinalibus tenuissimis in areis planulatis per undulatis; anfractibus, infra medium ungulatis, ultra angulum plano-concavis, prope suturas marginatis.*"

Axis, $1\frac{1}{2}$ inch.

Locality. Red Crag, Felixstowe.

There is some confusion respecting the name of this species. Lamarck described two species as *interruptus,* one a recent and very distinct shell, the other a fossil for which he adopted the specific name of (*Murex*) *interruptus,* referring it to the *Murex interruptus* figured and described by Brocchi in 1814; but a shell named *Murex interruptus* had been described by Pilkington in ' Trans. Linn. Soc.,' for 1804, vol. vii, T. 11, f. 5 (and also figured in ' Min. Conch.,' T. 304), which takes precedence and is entitled to that specific name. I have therefore adopted the above specific name of *turris* for the fossil from the Red Crag, Pilkington's species being a British Lower Tertiary form, and quite distinct from our present shell which is a Bolderberg and Italian species.

Bellardi has represented two shells under the name of *Pleurotoma interrupta,* considering them only as varieties of the same species, and the specimen from the Red Crag at Waldringfield, figured in my first Supplement, T. V., f. 1, seems to correspond with his variety C, given in fig. 11 of Tab. I of his work, while the present shell corresponds with his fig. 16 of the same plate. [Our specimen therefore seems to have got into the Red Crag from some bed corresponding to those of the Bolderberg.—ED.]

I found also among my siftings in the Red Crag at Felixstowe a considerable portion of a specimen of a species belonging to this genus with very distinct ornamental ridges or costæ which appears to correspond or at least to approach nearer to *Pleurotoma abnormis* of F. Edwards, ' Eocene Mollusca,' p. 294, Tab. XXX, fig. 14, *a. b.*, than to any other species I have compared it with. This being a London Clay species it may have come into the Red Crag with the *Rostellariæ* which I have figured. I also obtained a fragment of what seems to be *Pleurotoma Gastaldi*, Bellardi, Tab. II, fig. 19, but neither of them being in a condition to allow of correct determination I have not thought it worth while to have them figured.

Fig. 5 of Tab. I, represents one of two small specimens kindly sent to me by Dr. Reed with the name of *Pleurotoma gracilior*, A. Bell, from the Red Crag of Walton Naze affixed to it. These appear to have lost their outer coating, but are the same as the shell represented in fig. 12 of Tab. VII of vol. i of ' Crag. Moll.,' under the name *lævigata*, Phil., and which at p. 41 of my first Supp., is referred to *P. tenuistriata,* A. Bell. One of them has the upper whorls destroyed, but the other has all the whorls perfect and so peculiar that I have had it represented. It shows not only an obtuse apical region, but the first volutions are wholly different from the more cylindrical volutions of the rest of the shell.

PLEUROTOMA NEBULA, *Mont.*   3rd Supp., Tab. I, fig. 7.

FUSUS ? NEBULA, *S. Wood*		Catal. Mag. Nat. Hist., p. 541, 1842.
CLAVATULA —	—	Crag Moll., vol. i, p. 60, tab. vii, fig. 10, 1848.
PLEUROTOMA —	—	1st Supplement, p. 45, tab. vii, fig. 7, 1872.
MANGELIA —	*Forb. & Hanl.*	Brit. Moll., vol. iii, p. 476, pl. 114, figs. 7—9, 1853.

Although I have already given two figures of the Crag shell under the above specific name, they neither of them show a satisfactory representation of this long known species, and I have therefore determined to give another of a specimen in a more perfect condition from the cabinet of Mr. Robert Bell, which has retained some of its spiral striæ.

PLEUROTOMA HARPULA, *Brocchi.* 3rd Suppt., Tab. 1, fig. 4.

MUREX HARPULA, *Brocchi.* Conch. foss. Subap., p. 421, tab. viii, fig. 12, 1814.
PLEUROTOMA — *Phil.* En. Moll. Sic., vol. ii, p. 173, 1844.
FUSUS — *Risso.* Hist. Nat. Europe Mérid., vol. iv, p. 208, 1826.
RAPHITOMA — *Bellardi.* Monog. de Pleurot., p. 101, No. 22, 1847.

Axis, $\frac{9}{20}$ of an inch.
Locality.—Boyton.

A single specimen has been sent to me for examination and illustration by Mr. Robt. Bell, with Brocchi's specific name attached, and in this assignment I quite coincide. It appears in shape to be intermediate between *Fusus* and *Pleurotoma*, but probably only doubtfully to be entitled to the above generic position, as it seems quite destitute of the " side slit " of that genus. Our shell may be described in the words of Brocchi, viz.: " Testa turrita, longitudinaliter costata costis (8—9) tenuis, spiraliter striatis, interstitiis lævigatis, anfractibus convexiusculis, apertura ovata ; cauda brevissima aperta.

RAPHITOMA SUBMARGINATA, *Bellardi.* 3rd Suppt., Tab. 1, fig. 2.

PLEUROTOMA SUB-MARGINATA, *Bonelli.* Cat. Mus., *fide* Bellardi.
RHAPHITOMA — *Bellardi.* Monog. Pleurot. foss., p. 95, tab. iv, fig. 20, 1847.

Axis, $\frac{6}{10}$ of an inch.
Locality.—Red Crag, Felixstowe.

A single specimen, but unfortunately not quite in perfection, has been found in my siftings of the Red Crag material at Felixstowe, and I have referred it as above, but my dependence for so doing has been upon the description and figure by Bellardi, not having a specimen of the Italian fossil for comparison. My shell appears to be somewhat intermediate between this and *R. plicatella*, but I have no doubt that it is one of the very large group of fossil shells varying in some trifling degree only which connect the genus *Pleurotoma* and *Fusus*, and for which I believe nearly twenty generic divisions have been proposed. My shell is not far removed from *Murex vulpeculus*, Brocchi, and *Pleurotoma Maggiori*, Phil., forms. which, I think, might without any impropriety be specifically united. My shell measures six-tenths of an inch in length, and two-tenths in its diameter, without any ridges or folds upon the columella, or any denticulations on the inside of the outer lip; but this may be from its not having arrived at maturity. There are traces of spiral striæ, but the specimen has had its surface much eroded, and when perfect it was probably fully covered. It has about a dozen costulæ or riblets on the last volution. [The specimen appears to me to be a derivative.—ED.]

COLUMBELLA ERYTHROSTOMA? *Bonanni.* 3rd Suppt., Tab. 1, fig. 10 *a, b.*

COLUMBELLA ERYTHROSTOMA, *Bon. Fide* Bellardi Monog. delle Columbelle foss. del Piedmonte, p. 9, fig. 4, 1848.

Spec, char.—" *Testa turrito-elongata, turgidula, anfractibus lævibus, convexiusculis ; ultimo magno : apertura dilatato-elongata, labro subarcuato, subvaricoso ; columella adnata, regulariter et numerose rugosa ; rugis brevibus externis.*"—Bellardi.

Locality.   Red Crag, Butley.

The above figures represent specimens found by myself some years ago, in the Red Crag of Butley, which I have hitherto left unnoticed, regarding them merely as specimens of *C. sulcata,* Sowerby, derived from an older part of the Red Crag, and worn smooth in consequence, that species being abundant at Walton, and variable in length; one figured in Supplement to Crag Moll., p. 9, Tab. 11, f. 16, measuring one inch and five-eighths, while another is less than three-quarters of an inch, both of them being full-grown, and belonging, I believe, to the same species.

The specimens now figured are quite smooth, a character agreeing with that which Bellardi has given for the Italian fossil *erythrostoma,* which is described as " *anfractibus lævibus ;* " but if my specimens have been derived from an anterior Red Crag bed, they may have lost the spiral striæ from either decortication or abrasion, and so be, as I originally supposed them to be, merely worn specimens of *C. sulcata.*   Mr. A. Bell gives three specimens of this genus from what he terms the Middle and Upper Crag, viz. *C. sulcata, C. abbreviata,* and *C. Borsoni ;* and another is added in Prof. Prestwich's catalogue of mollusca from the Red Crag, viz. *C. scripta.*   In my original work, and in the supplements thereto, I have figured several different forms of what appear all to be *C. sulcata ;* and as two specimens, which had been furnished him by Mr. A. Bell, under the name of *Columbella abbreviata,* have been kindly sent to me by Dr. Reed, I have figured one of them (Tab. 1, fig. 6), in order that a representation of the shell, on the strength of which this name of *abbreviata* has been introduced into the list of Red Crag Mollusca, may appear. The shorter of the two specimens which I have figured under the name of *erythrostoma* (fig. 10A), agrees with this *abbreviata,* but is smooth.

LACUNA (MEDORIA) TEREBELLATA, *Nyst.*

> MELANIA TEREBELLATA, *Nyst.*   Coq. foss. de Belge, p. 413, pl. xxxviii, fig. 12, 1843.
> PALUDESTRINA   —   *S. Wood.*   Crag Moll., vol. i, p. 109, tab. xii, fig. 7, 1848.
> EULIMENE   —   —   1st Supplement, p. 65, 1872.

This shell was figured by myself in the ' Crag Moll.' under the generic name of *Paludestrina.* In my first Supplement I, in my perplexity, grouped it in a new genus, in which I proposed to embrace another crag shell, viz., *Eulimene.* It is not, I think, either a freshwater or an estuarine shell, neither does it belong either to *Paludina* or to *Littorina.*

In the Red Crag at Felixstowe I have lately obtained more than a hundred specimens, varying in the length of axis from an eighth of an inch to upwards of five eighths, every one of which is in a mutilated condition, but all belonging to this species (whatever it may be) ; and every one has, more or less, its umbilicus (lacuna), covered over, by apparently, an extension of the left lip of the shell. This extremely mutilated condition evidently indicates that the specimens have been introduced into the Red Crag both at Walton and elsewhere from some older bed, but I have not been able to trace whence. They are very thick and strong shells, more so than any freshwater species in this country.

[The shell is described by M. Nyst, in his ' Coq. foss. de Belge,' as occurring at Antwerp and Calloo, and as being rare, but he does not there specify in what division of the Upper Tertiaries at these places the shell is found. In his ' Listes des Fossiles des divers Etages,' p. 424, however, he gives it from the Crag jaune (or uppermost crag) only. I do not find it in any of the lists given by M. Vanden Broeck, in his ' Esquisse Geologique,' for the different horizons which he seeks to establish of the beds at, and in the neighbourhood of Antwerp.—ED.]

In the ' Crag Moll.,' vol. i, p. 108, Tab. XI, fig. 2 *a, b*, is figured and described a shell from Bramerton, under the name of *Paludestrina subumbilicata*, which may, I now think, be regarded as the ancestor of the living *ventrosa*, and it is there stated that in my cabinet was one specimen from the Cor. Crag, the identity of which was given as doubtful in consequence of the Bramerton shell (*subumbilicata* or *ventrosa*) being generally considered a freshwater or estuarine inhabitant. This species, however, as well as *ulvæ*, is capable of living where the water is not quite fresh, and I have lately found in the purely marine Red Crag of Felixstowe a few specimens which appear to me undistinguishable either from the Bramerton shell, or from the living species, called by the British Conchologists *Hydrobia ventrosa.* If we may depend upon figures and descriptions, there are several continental shells with different names (both generic and specific) which cannot be separated from the Crag and recent shell above referred to, but of these some are given as fossils from deposits that are said to be purely of freshwater origin, while others are given as from beds of purely marine origin. This species so closely resembles some of those of *Rissoa*, that I do not know any character in the testaceous part by which it can be separated from that genus.

NODOSTOMA ORNATA, *S. Wood.* 'Crag Moll.,' vol. i, p. 87, Tab. IX, fig. 6, as *Odostomia simillima;* 1st Sup., p. 64, as *O. ornata;* 3rd Sup., Tab. I, fig. 13.

Locality.—Cor. Crag, Sutton.

This pretty little shell was figured and described in the 'Crag Moll.," under the name of *Odostomia simillima*, and was assigned to Montagu's species *simillimus*, which I now consider was erroneous; and in my first Supplement I assigned it as distinct, and gave it the name *ornata*. The obscure tooth, stated in my first volume (p. 87) as present upon the columella, is, I find, only a fragment of sand adhering to the columella, while the aperture is more elongately ovate than in *Odostomia*, and of quite a different form from that in *Chemnitzia*. My specimens were very few and somewhat variable, but the species, I think, cannot be placed in the genus *Odostomia*, being apparently intermediate between that genus and *Eulima*. I therefore propose to call it *Nodostoma*[1] from its evident relationship with *Odostomia*, but separated from it by its toothless character.

The shell described by Montague is considered by the authors of ' Brit. Moll.,' as well as by the author of ' Brit. Conch.,' to have been " a bleached and worn specimen " of *Chemnitzia rufa*, Phil., and doubtfully British. The present figure is taken from a single specimen that I have recently found, the shell being extremely rare.

NODOSTOMA EULIMELLOIDES, *S. Wood.* 3rd Sup., Tab. I, fig. 14.

Locality.—Cor. Crag, Sutton.

[Of the specimen figured as above, a sketch was made by my father for his intended plate under this name; and he appears to have intended to give it as a second species of his new genus, *Nodostoma*, but he has left no other MS. respecting it beyond the above specific name of *eulimelloides*. I have compared it with all the species of *Eulima* described by him from the Crag, and it agrees with none satisfactorily. It comes nearest to *Eulima glabella*, but the form of the mouth differs, the whorls are more cylindrical,

[1] Νωδος, toothless, and στομα, mouth.

and the suture is deeper or more marked. The surface is smooth and without any ornament. Though imperfect by the loss of the upper whorls, the specimen is otherwise in good preservation, and shows these distinguishing characters clearly.—ED.]

MENESTHO ? SUTTONENSIS, *S. Wood.* 3rd. Sup., Tab. I, fig. 11.

Locality.—Cor. Crag, Sutton.

The above figure represents a small shell found by myself some years ago and retained until now in the hope of obtaining a better specimen. I have referred it to the genus *Menestho,* as to which I have made some remark at p. 56 of my first Supplement.

My shell is unfortunately not quite perfect, the outer lip being slightly broken, but it much resembles the opening of *Rissoa* or *Odostomia.* The specimen is covered with four rather coarse spiral lines and depressions on the lower whorl, and three on the next above this, but probably it may not be a full-grown shell. The nearest figure to which I have been at all able to refer it (approximately) is a very small shell, described by Isaac Lea in his contributions to 'Geology,' pl. iv, fig. 84, under the name of *Pasithea sulcata,* but, judging from this figure, my shell is distinct. Lea gives no less than nine species under that generic name, several of them differing materially in characters that it would be difficult to collect into one genus, and he does not specify which of these he regards as the type of his genus *Pasithea,* so that I am unable to adopt that genus for my present species.

ODOSTOMIA REEVEI, *S. Wood,* 3rd Supp., Tab. I, fig. 12.

Locality. Fluvio-marine Crag, Bramerton.

The above figure represents a specimen of the above-named genus sent to me by Mr. Jas. Reeve, of the Norwich museum and found by him at Bramerton in the bed which yielded the specimens of *Cerithium derivatum* and *Odostomia derivata* described in the 'Second Supplement to the Crag Moll.' (pp. 39—40). The nearest species to which I can compare it is *O. dubia,* Jeff., but it differs sufficiently, I think, to be considered distinct, at least as much so as several of our so-called British species. The shell is somewhat thick and free from striæ of any kind, the aperture measures half the length of the entire shell, and is of a very ovate form, the base of it being contracted more than usual in any species of this genus. The shell is rather larger than any of my specimens from the Cor. Crag, with the exception of *O. conoidea* and *O. turrita,* which have eleven volutions while the present shell has not more than four, or perhaps five.

In the 'British Mollusca," and in the 'British Conchology,' there are more than twenty *Odostomiæ* described as distinct species, each with very slight differences of character; but whether they are all specifically distinct is perhaps questionable. The Authors of 'British Mollusca,' vol. iii, p. 260, justly say: " The species are difficult to distinguish and very critical." I have figured several so-called species under this generic name and I have in most cases assigned them from the figures and descriptions of these Authors, and of the Author of 'British Conchology,' as they had better means for determination than I have had.

[The specimen figured is probably one which has been carried into the fluvio-marine Crag from the same bed as that which supplied *Cerithium derivatum* and *Odostomia derivata.*—ED.].

[The following description of a new species and some remarks as to the bed at Boyton, in which it occurred, have been kindly supplied by Mr. Robert Bell.—ED.]

[MARGARITA CRASSI-STRIATA, *Robt. Bell.*   3rd Sup., Tab. 1, fig. 15.

Locality. Boyton.

Shell small, very solid, somewhat conical; whorls five; suture deep, each volution having four or five thick revolving ridges with traces of fine intermediate ridges. These are crossed by prominent lines of growth, giving them a slightly crenulated appearance. The base is, like the whorls, rounded and strongly ridged, with a very small umbilicus. Mouth rounded, with an obscure tooth or fold near the base of the columellar lip.

The species which seems nearest to it is *Margarita cinerea*, Couthuoy, but it differs in having much stronger ridges, especially at the base, and a smaller umbilicus. The upper whorls also do not seem to have that lattice-like appearance which is present in well-preserved specimens of *M. cinerea*.

It is difficult to indicate which formation this shell belongs to. The section of Crag worked at Boyton can seldom be seen, being an excavation close to the Butley River, and mostly from three to six feet under water, the coprolite diggers standing in the water when at work, and scooping up the sand from the bottom of the trench; but from what I have been able to observe, and from an examination of a large number of species found there, the formation seems to range from the fossiliferous beds of the Coralline (Zone d. of Prestwich's section in his paper on the " Crag Beds of Suffolk and Norfolk," 'Quart. Journ. Geol. Soc.,' vol. xxvii, p. 121,) up to the middle portion of the Red Crag. Probably some of the beds have been reconstructed from the wearing away of the Upper Coralline strata on the other side of the river, although a bed of the larger bivalves

Astarte, Cardita, &c.) was seen some few years ago *in situ* at the base of the excavation, in a part now filled in, and I have obtained many double shells from there exactly answering to those found in the pits at Broom Hill, Sudbourn, and at Sutton. There seems also to be an admixture of shells from some formation with which we are unacquainted in England (most probably the Belgian Crag) as several species have been found here that have not been detected in any other Crag bed (*Fusus Waelii, Murex Reedii,* &c.). The Red Crag element is, however, sufficiently prevalent, and such shells as *Trophon scalariformis, T. muricatus,* and especially *Nassa reticosa,* are particularly abundant.[1] The specimen of *Amaura candida* mentioned in the column of remarks in the list of Mollusca given in the first 'Supplement to the Crag Mollusca,' as found at Boyton, came, I believe, from Butley, *i. e.* from the same locality as the specimen figured in Tab. I, fig. 3, of that Supplement. Robt. Bell.]

BIVALVIA.

SILIQUARIA PARVA, *Speyer.* 3rd Sup., Tab. I, figs. 16 *a—b.*

> SILIQUARIA PARVA, *Speyer.* Ober.-Oligoc. Tertiar. Detmold., p. 33, tab. iv, fig. 2 *a, b,*
> Palæontographica, Band xvi, 1869.

Spec. Char. " *Testa parva tenuissima, oblonga, antice brevis, postice producta, utrinque æqualiter rotundata, lævigata, nitida ; cardo subumbone parvulo fossula plana instructus, dente unico munitus. Nymphæ breves angustæ.*" Speyer.

Locality. Bramerton.

Two fragmentary specimens of a small bivalve were sent to me by Mr. Jas. Reeve (as mentioned in my second Supplement, p. 40), which I thought were too small and imperfect to be represented, but as they appear to be indicative of the presence in Norfolk of an older formation than the one in which they have been found, I think it desirable to figure them, imperfect as they are. The hinge has a prominent fulcrum for the support of its external connector, the central tooth large, prominent, and obtuse, being immediately before it and under the umbo ; and there is a depression in the corresponding valve for its reception[2] similar to the hinge furniture of *Saxicava,* which it much resembles, as it does also the shells of *Sphenia,* but there appears, I think, sufficient difference to

---

[1] [See also footnote to p. 3 of Second Supplement as to this Boyton bed, the information quoted there having been obtained from Mr. Alfred Bell. From that it would appear that the bed containing *Astarte* and *Cardita* was part of the lowest portion of the Coralline Crag, and was overlain by some Red Crag ; the shells of both formations becoming thus intermingled in the working.

[2] The engraver has not been successful in delineating the character of the hinge in either valve. The generic name *Siliquaria* is used here from Speyer, but it is that also of a vermiform shell.—ED.]

justify a generic distinction. The hinge more resembles that of the latter shell, but that species (*Sphenia*) has an internal connector. The name of *Siliquaria* (of Schumacher), as given to the Oligocene shell by Dr. Speyer, is, I think, sufficient to guide us in our future determination, for although I have many hundreds of specimens of *Saxicava* of small size from the Coralline Crag, I have nothing that will fairly correspond with the present shell.

[The specimens have probably got into the Fluvio-marine Crag of Norfolk from the same formation there which supplied those of *Cerithium derivatum, Odostomia derivata,* and *Odostomia Reevii.*—ED.]

CARDIUM ECHINATUM, *Linn.* Crag Moll., vol. ii, p. 152.

As stated at p. 152 of my second volume this species has very rarely occurred in the Crag, but a specimen has lately been found at Felixstowe by Mr. W. E. Hardy, of Park Crescent, Stockwell, which was sent to me for verification, and it is similar to the one (now in the British Museum) figured in the 'Crag Moll.,' vol. ii, p. 152, Tab. XIV, fig. 3. It belongs probably to the variety called *ovata* by Dr. Jeffreys in 'Brit. Conch.,' vol. ii, p. 271, and described by him as having the "ribs sharp." The Crag shell has triangular ribs (unlike the common recent species, on which the ribs are quadrate), with spines in a slight depression down the centre of these. The species is very rare in my collection, I having found no other specimen than the one I gave to the British Museum. This specimen is in good preservation with the exception of having lost all its spines. I have a shell from the Sicilian beds which it more resembles, with sharp angular ribs covered with broad spatulate imbricated spines, but Mr. Hardy's specimen, though well preserved otherwise, has lost all. I do not know whether this Sicilian fossil has ever been figured.

PECTEN DISPARATUS, *S. Wood.* 3rd Suppt., Tab. I, fig. 17.

Locality. Red Crag, Waldringfield.

The shell as above represented has been sent to me by Mr. R. Bell, but without a name, and I know not to what published species it can be justly referred. I thought at first that it might be one of the many varieties of that variable shell *P. Danicus* (*septem radiatus*), but I have not been able to find one precisely similar in character; and although there is much resemblance to two or three other species, I have not been able to assign it satisfactorily to any one. I have therefore given to it provisionally the above name. It is somewhat similar to *P. multicarinatus*, Lam., figured and described by the late Dr. Deshayes, 'Descr. de Coq. foss. des Env. de Par.,' p. 307, Pl. XLII, figs. 17, 18, 19, but that is not quite so large a shell, and is said to be from Parnes, in the upper portion of the Paris Eocene. It differs essentially from *P. duplicatus*, on which the ribs are nearly uniform in size. Our shell is nearly orbicular, covered with ten or twelve large and slightly prominent convex rays, upon which, and also between them are three smaller rays, and between each of these is an alternate smaller one, so that between each of the most prominent there are seven smaller. All of these are ornamented with sharp imbrications, and the shell has unequal auricles, which in our specimen are not quite perfect; but there are indications of these being of large size in the perfect shell. In the interior of this valve, which is the right one, there are eight or nine furrows corresponding to the elevation of the prominences of the larger ribs. The muscle mark is not very distinct. This specimen, is, in all probability, a derivative from an older formation.

OBSERVATIONS AS TO THE SUCCESSIVE FORMATION OF THE BEDS FORMING THE APPARENTLY HOMOGENEOUS AND SYNCHRONOUS MASS OF "RED CRAG," AND THE ILLUSORY CHARACTER OF THE EVIDENCE AFFORDED BY PART OF THE ORGANIC REMAINS IN THEM.

HAVING in a previous portion of my work on the Crag Mollusca expressed my opinion of the distinctive character of the beds at Walton Naze from the main portion of the Red Crag, and of their older age, I took the opportunity of a few months' stay at Felixstowe in 1879-80 to thoroughly sift and search a large quantity of the Red Crag there, to ascertain not merely what species of Mollusca could be detected in it, but also the general condition in which the remains of these were preserved, so as to compare them with those at the Walton Naze locality, with which, from many visits to that place in the earlier years of my study of the subject, I was very familiar.

The following list is the result of that investigation; and in it I have affixed to those species which appear to me to have come into the Red Crag of Felixstowe only by derivation from beds older than the Red Crag (including those of the Coralline Crag,) the letter D, while to those which appear to me to have come only by derivation from earlier beds of Red Crag age, such as that at Walton Naze, I have affixed the letter W, the exclusively fragmentary condition of some species being indicated by the letter F.

REMAINS OF MOLLUSCA[1] FOUND IN THE CRAG OF FELIXSTOWE.

Gasteropoda.

Cypræa Europea, *Mont.*
— avellana, *J. Sow.*, W.
Voluta Lamberti, *J. Sow.*, F, D, W.
Terebra inversa, *Nyst*, F, D.
— canalis, *S. Wood*, F, D.
Columbella sulcata, *J. Sow.*, F, W.
Cassidaria bicatenata, *J. Sow.*, F, D.
Nassa granulata, *J. Sow.*
— incrassata, *Müll.*
— consociata, *S. Wood*, F, D.
— propinqua, *J. Sow.*
— pygmœa, *Lam.*
— labiosa, *J. Sow.*, F, D.
— reticosa, *J. Sow.*, W. and mostly F. or imperfect.
Rostellaria lucida, *J. Sow.*, F, D.
gracilenta, *S. Wood*, F, D.
Buccinum Dalei, *J. Sow.*
— undatum, *Linn.*
Purpura lapillus, *Linn.*
— incrassata, *J. Sow.*
— tetragona, *J. Sow.*, F, W.
Murex tortuosus, *J. Sow.*, F, D.
Trophon antiquus, *Linn.*
— — id. *var.* contrarius.
— alveolatus, *J. Sow.*, F, D.
— costifer, *Nyst*, F, W.
— altus, *S. Wood.*
— gracilis, *Dacosta.*
— muricatus, *Mont.*
— — id. *var.* exossus.
— Olavii, *Beck.*
— scalariformis, *Gould.*
Pleurotoma interrupta, *Broc.*, F, D.
— turricula, *Mont.*

Pleurotoma Trevelyana, *Turt.*
— scalaris, *Möll* (one specimen full size and perfect).
— nebula, *Mont.*
— costata, *Dacosta.*
Cancellaria scalaroides, *S. Wood*, F, D.
— (Admete) viridula, *Fab* (one specimen broken).
Cerithium tricinctum, *Broc.*, F.
— variculosum, *Nyst* (one whirl only), F, W.
— granosum, *S. Wood?* F, W.
Aporrhais pespelicani, *Linn.*, F, D. (very worn fragments).
Turritella incrassata, *J. Sow.*, F. and mostly D.
Scalaria funiculus, *S. Wood*, F, D.
— foliacea, *J. Sow.*, F, D.
Chemnitzia internodula, *S. Wood.*
Eulima intermedia, *Cant.*, D and W?
Eulimene pendula, *S. Wood.*
Lacuna (Eulimene) terebellata, *Nyst.*, D.
Rissoa curticostata, *S. Wood.*
Littorina littorea, *Linn.*
Natica catena, *Da Costa.*
— catenoides? *S. Wood.*
— clausa, *Brod.* and *Sow.*
— hemiclausa, *J. Sow.*
— multipunctata, *S. Wood.*
Vermetus intortus, *Lam.*, D?
Trochus cinerarius, *Linn.*, W? (the specimens are all slightly mutilated).
— Montacuti, *W. Wood.*
— tumidus, *Mont.*
— zizyphinus, *Linn.*, F, D.

[1] The absence of a capital letter after the name of a species means that that species is not derivative.

Fissurella Græca, *Linn.*
Emarginula fissura, *Linn.*
Calyptrœa Chinensis, *Linn.*
Capulus Ungaricus, *Linn.*
Tectura virginea, *Möll.*

Dentalium dentalis, *Linn.*, F, D.
— entalis, *Linn.*, D? (worn).
Ringicula buccinea, *Broc.*, F, D.
Bulla cylindracea, *Penn.*, F.
Melampus pyramidalis, *J. Sow.*

Bivalvia.

Anomia, *sp.*
Ostrea, *sp.*
Pecten maximus, *Linn.*, F, D.
— opercularis, *Linn.*
— pusio, *Penn.*
Lima exilis, *S. Wood*, F, D, W?
Mytilus edulis, *Linn.*, F.
Arca lactea, *Linn.*
Pectunculus glycimeris, *Linn.*
— subobliquus, *S. Wood*, W.
— pilosus, *Linn.*, D.
Nucula lævigata, *J. Sow.*
— Cobboldiæ, *J. Sow.*
— nucleus, *Linn.*
Leda oblongoides, *S. Wood.*
Lucina borealis, *Linn.*
Diplodonta astartea, *Nyst.*
Cardita senilis, *Lam.*, D.
— scalaris, *Leathes.*
— chamæformis, *Leathes*, D (worn).
— corbis, *Phil.*
Cardium angustatum, *J. Sow.*
— decorticatum, *S. Wood*, D.
— edule, *Linn.*
— echinatum, *Linn.*
— Parkinsoni, *J. Sow.*
— venustum? *S. Wood.*
Astarte Basterotii, *de la Jonkaire*, F, D.
— Burtinii, *de la Jonkaire*, D.
— crebrilirata, *S. Wood.*

Astarte incrassata, *Broc.*, D.
— obliquata, *J. Sow.*
— Omalii, *de la Jonk.*, F, D.
— compressa? *Mont.*
Woodia digitaria, *Linn.*
Cyprina islandica, *Linn.*, F.
Venus casina, *Linn.*, F, D.
— fasciata, *Da Costa.*
Cytherea chione, *Linn.*, F, D.
— rudis, *Poli.*
Artemis lentiformis, *J. Sow.*, F, W.
Tapes pullastra, *W. Wood*, F.
— virgineus? *Linn.*, F.
Gastrana laminosa, *J. Sow.*, F, D.
Donax politus, *Poli*, F, D?
Psammobia, *sp.*, F, D.
Tellina obliqua, *J. Sow.*
— prætenuis, *Leathes.*
Mactra arcuata, *J. Sow.*
— ovalis, *J. Sow.*
Solen siliqua, *Linn.*, F.
— ensis, *Linn.*, F.
Corbula striata, *Walk.*
Corbulomya complanata, *J. Sow.*, W?
Saxicava arctica, *Linn.*
Panopea Faujasii, *Men de la Groye*, F, D.
Mya arenaria, *Linn.*, mostly F.
Pholas crispata. *Linn.*, F.
— cylindrica, *J. Sow.*, F, W?
Gastrochæna dubia, *Penn*, F, D.

[Mr. Robert Bell, who has of late years very assiduously searched the Walton beds, as well as examined several collections made by others from that locality, has kindly furnished the following list of all the molluscan remains which he has been able to detect there, beyond those given in the column for that place in my father's lists in the first Supplement to his work. The species to which an asterisk is affixed are additions to the mollusca of the Upper Tertiaries of the east of England, given in the previous part of this work, and are inserted solely on the authority of Mr. Bell.

Gasteropoda.

Erato lævis, *Don.*

Nassa labiosa, *J. Sow.*

— propinqua, *J. Sow.*

Buccinum undatum, *Linn*

Trophon consocialis, *S. Wood* (one specimen only, much worn, and probably derivative).

— gracilis, *Da Costa.*

— scalariformis, *Gould.*

Pleurotoma linearis, *Mont.*

— turrifera, *Nyst.*

— nebula, *Mont.*

— rufa ? *Mont.*

Turritella planispira, *S. Wood.*

Chemnitzia communis,* ? *Risso.* (perhaps only a short form of *C. internodula.*)

Eulima subulata, *Don.*

Odostomia acuta, *Jeff.**

Natica catena, *Da Costa.*

— clausa, *Brod.* and *Sow.* (affinis. of *Gmel.*)

— varians, *Dujardin.*

Vermetus intortus, *Lam.*

Trochus formosus, *Forbes.*

— multigranus, *S. Wood.*

— Adansoni, *Payr.*

— tumidus, *Mont.*

— Kicksii, *Nyst.*

— Montacuti, *W. Wood.*

— zizyphinus, *Linn.*

Emarginula crassa, *J. Sow.*

Tectura virginea, *Müll.*

Dentalium dentalis, *Linn.*

— rectum, *Linn.*

Actæon subulatus, *S. Wood.*

— tornatilis, *Linn.*

Bivalvia.

Mytilus edulis, *Linn.*

Modiola phaseolina? *Phil.*

Nucula nucleus, *Linn.*

— Cobboldiæ, *J. Sow.* ?[1]

Nucula tenuis? *Mont.*

Cardita senilis, *Lam.*

Cardium fasciatum, *Mont.*

Cardium strigilliferum, *S. Wood.*

[1] My father collected extensively at Walton at intervals during forty years, and Mr. Robert Bell also very assiduously for many years past, without either of them having met there with the slightest trace of this shell, so common in the later part of the Red Crag; but Mr. Bell has lately met with a single worn valve in the collection made from Walton by Mr. Greenhill, of Vermont College, Clapton, on the authority of which the shell is inserted with a note of interrogation in the above list.

Cardium pinnatulum, *Con.* (nodosulum).
Astarte Galeotii, *Nyst.*
— Forbesii, *S. Wood.*
Circe minima, *Mont.*
Abra prismatica, *Mont.*
Mactra glauca, *Born.*

Tellina obliqua, *J. Sow.* (a fragment only by Mr. Bell, another fragment by Mr. Hy. Norton of Norwich, and a single valve by Mr. Greenhill.)
Mya arenaria, *Linn.*

The contrast thus shown by the Crag of Felixstowe to that at Walton Naze (seven miles distant from it) is very striking. At the former place such species as *Trophon costifer*, and *Nassa reticosa*, among Gasteropods, which abound at Walton, and are there preserved in the most perfect condition, are, though abundant, scarcely to be found unmutilated; and such very few examples of them as do occur but little broken are all more or less worn. Among the Bivalvia one of the most abundant shells at Walton, *Artemis lentiformis,* and which at that place is almost always perfect (though generally with valves detached), is, though very abundant, *invariably* in fragments at Felixstowe. That this fragmentary condition at Felixstowe can only arise from the presence of the shell in the Crag there being due to derivation from the destruction of anterior accumulations, is shown by the fact that while *A. lentiformis,* which is thus in fragments is a strong shell, the thin and fragile shell, *Tellina prætenuis* (a species unknown from the Walton bed but in tolerable abundance at Felixstowe) occurs almost always perfect. It is, in my opinion, abundantly clear that during the time which elapsed between the accumulation of the Walton beds of Red Crag and their destruction and re-accumulation to form the Red Crag of Felixstowe, such shells as *Trophon costifer, Nassa reticosa,* and *Artemis lentiformis,* as well perhaps as some others had ceased to live in the Red Crag sea; and that other shells such as the dextral form of *Trophon antiquus, Leda oblongoides, Tellina prætenuis,* to which might have been added *Nucula Cobboldiæ,* but for the solitary and somewhat uncertain occurrence mentioned in the footnote on p. 16, (all of these being species which endured into the early Glacial sea,) and probably some others which might be mentioned, had been introduced into it. Moreover, the extremely profuse shell of all the rest of the Red Crag and of the Lower Glacial sands, *Tellina obliqua,* but which had lived in the Coralline Crag sea, was during the Walton accumulation so scarce in the Red Crag sea that only a single valve of it and two fragments (by three separate collectors) have been detected there.

In the Red Crag of Butley the change becomes further marked, both by the greater frequency of these later introductions, and by the presence of arctic species, which have not yet been detected in the Crag of Essex or of the more southern part of Suffolk, the Upper Beds of the Red Crag having either been removed from, or else having never been formed in, that part of Suffolk.

The changes which led to the peculiar and exceptionally perplexing features thus presented by the beds of the Red Crag of England, with their large admixture of false

3

evidence afforded by derivations from beds anterior to that Crag, to a smaller extent also by derivations from earlier beds of Red Crag age, appear briefly to have been these.

At the incoming of marine conditions over part of England after the long interval of terrestrial conditions which had endured since the elevation and denudation of the Oligocene sea-bed, and when several of the tropical genera of Mollusca characteristic of the older tertiary time still lived in the sea of our latitudes, the older Pliocene submergence seems to have extended from the north of Belgium, over the south-east of England, and in that way formed a strait, connecting the North Sea with an arm from the Atlantic which extended over Touraine.[1] The evidences of the oldest accumulations of this strait which remain in England are probably some sands on the Chalk Downs between Maidstone and Dover, and (I think it likely) also an outspread of shingle along the strait's northern shore, of which patches remain on the Lower Bagshot outliers of South Essex, and of the Isle of Sheppy,[2] and sweep over the edges of some of these on to the uppermost beds of the London Clay there, as well as of a patch of the same shingle crowning the middle part of the London Clay on Shooters Hill, in north-west Kent, and possibly some others on the chalk of North Surrey, near Caterham. Changes took place in the distribution of the land and water of this strait, and the Coralline Crag ensued.   Except over a part of Belgium, and (deeply buried under more recent beds) probably a part of Holland also, the oldest beds of this Pliocene Strait have been almost entirely removed by the later action of the sea, and numerous remains of the marine animals, both vertebrate and invertebrate, which were entombed in them have, in consequence, got into the Red Crag, particularly the nodule bed at its base. Remnants of the Coralline Crag, however, remain near each extremity of this Strait, viz. in Normandy near the one, and in Suffolk near the other end, besides a more general

[1] The French geologists still apply the term " Miocene" to the Faluns of Maine et Loire and of Touraine, although these Faluns appear to be coeval with beds in Belgium to which several of the geologists of that country apply the term "Pliocene," insisting that the "Miocene," *i. e.* the marine equivalent for the terrestrial interval between the "Oligocene" and the oldest "Pliocene," is not represented by any marine deposits there.   To avoid as much as possible adding to this confusion, especially as the oldest part of the English Crag—the Coralline—is clearly "Pliocene," I have avoided in the text the use of the word " Miocene."   The beds of Maine et Loire and of Touraine not only contain many shells of the Coralline Crag which do not appear to be survivors from the older Tertiary seas of England and France, but also living British shells, such as *Murex erinaceus,* which do not appear to have entered British seas until the time of the Red Crag, or, such as *Nassa reticulata,* even until the Glacial submergence.

[2] See 'Quart. Journ. Geol. Soc.,' vol. 24, p. 464, and bed No. viii, of the plate in vol. 36, p. 457. Prof. Prestwich, in a paper "On the Extension into Essex, Middlesex, and other inland counties, of the Mundesley and Westleton Beds," read before the Brit. Assoc. in 1881, appears to refer the shingle mentioned in the text as occurring on the Lower Bagshot outliers to the Lower Glacial pebbly sand (No. 6 of the beds described in the "Introduction" to the first Supplement to the Crag Mollusca); from which view, as well as from others in the same paper, I differ. My own view of the events which took place during the Newer Pliocene period in England is given in a memoir of which the first part is published in the 36th volume of the 'Quarterly Journal of the Geol. Soc.,' p. 457.

outspread in Belgium. By the gradual emergence of this strait the sea in Belgium and East Anglia, at the time represented by the Red Crag, *i.e.* the commencement of the Newer Pliocene period, had become separated by land from that in Normandy, but the molluscan remains which it has left in the latter country closely agree with those of the older portions of the Red Crag of East Anglia.[1] One of the results of this separation seems to have been to cause, on the English Coast of the North Sea, a great rise and fall of the tide over a very shallow and flat bottom. As this tide surged round the low island of Coralline Crag at Sutton, and also round the peninsula of the same Crag formed by the parishes of Sudbourn, Orford, and Aldboro' (the rest of the Coralline Crag, with some small exception, having been destroyed either during emergence by the sea which deposited it or by the inroad of the Red Crag water), it carried from that Crag a large quantity of its Molluscan remains which thus became mixed with the remains of the Mollusca then living in this sea, so that the banks of Red Crag, which were then accumulating in South Suffolk, became full of such derivatives, while the bed at Walton, being more distant from that island and peninsula, was left almost entirely destitute of organisms of this extraneous origin.

Formed under these conditions, and accumulated as banks or foreshores between high and low-water mark, as their peculiarly continuous highly oblique bedding attests, the marine beds of the Red Crag (with the exception of the latest or Chillesford beds of that formation, which accumulated during a slight depression of the area at the close of the Crag,) were continuously undergoing destruction and reaccumulation; and successive accumulations of them, formed between tide marks, may be seen in some sections laid up at the foreshore angle of bedding, one upon another. Thus the changes in the molluscan life of the North Sea, which from the approach of the glacial period were taking place during the Red Crag, have become obscured by the circumstance that the remains of mollusca which had died out (in that sea at least) were, in consequence of the destruction of these older banks, and the reaccumulation of the material of them in new banks of the same character and mode of deposit, mixed up with those of the mollusca still surviving there, and of some new forms which the change of climate, and probably distant geographical changes also, were bringing in; this mixed accumulation being further complicated by the introduction of molluscan remains from the Coralline Crag and still older formations.

---

[1] See 'Étude Geologique sur les Terrains Crétaces et Tertiares du Cotentin,' par. MM. Viellard and Dollfus, Caen, 1875, pp. 148—163. The material of these beds of the Cotentin referable to the Coralline Crag (*Conglomérat à térébratules*), of which Mr. Harmer brought me some from St. Georges de Bohon, near Carentan, appears undistinguishable, both in mineral character and included organisms, from the Upper Beds of the Coralline Crag, at Sudbourn.

I take this opportunity of correcting the representations given by Mr. Harmer and myself of the beds of the Crag district in the map, and sections which accompany the "Introduction" to the first Supplement to the Crag Mollusca in the volume of the Society for 1871, so far as subsequent observations have rendered necessary, as follows:

Owing to the obscurity existing where sand rests on sand, the Lower Glacial sand, No. 6 of the map, is not shown further south than the neighbourhood of Dunwich; and in the section (A) through the Red Crag area it is omitted altogether, and the Middle Glacial (No. 8) represented as resting throughout on the Red Crag. Residence in the district since 1873 has afforded me the means of a closer examination and comparison of pit sections there, and convinced me that this representation (which was mine only) was erroneous, and that the sand No. 6 is not only present, but is the principal formation in this area; for though it is mostly underlain by Red Crag, it in many places takes the place of this, and rests direct on the London Clay. Over the Red Crag, however, there is in some excavations a reddish-brown sand, soft, loamy, and destitute of the smallest fragment of shell, but in which sometimes masses of shelly crag are enveloped, and in which, in some rare instances, bands of ironstone containing casts of Red Crag shells also occur. This sand is merely the Red Crag from which the calcareous constituents have been carried away by dissolution in water, while the argillaceous and ferruginous constituents have been either left unaffected, or else redeposited in the undisturbed sandy mass. The difficulty, therefore, is to distinguish between this and the sand No. 6; for in South Suffolk the latter loses the shingly or pebbly character which enables it to be easily recognised in North East Suffolk and in Norfolk. Over the Red Crag area the sand No. 6 passes upwards by the mere substitution of argillaceous for arenaceous sediment into stratified brickearth, just as it does on the Cromer Coast and generally in North Norfolk, though from its geographical position in South Suffolk this brickearth has not there received that copious intermixture of chalk *débris* and chalk silt which along the Cromer Coast (where it is represented by the "Contorted Drift," bed No. 7 of the Map, &c.) forms its preponderating constituent, in proportion to the diminution in its distance from the Lincolnshire Chalkwold, from the degradation of which by the land ice during the earlier part of the Glacial period, when England was undergoing its great submergence, this *débris* and silt were derived; but thin layers of this *débris* are sometimes present in it in South Suffolk, as *e.g.* at Kesgrave. Neither has it been so disturbed by the action of grounding bergs as in North Norfolk, where the result of this action has obtained for it the name of "Contorted Drift;" nevertheless, it is sometimes contorted in Suffolk, as I observed in an excavation of it beneath the chalky clay on the Hasketon side of Woodbridge in 1874. Over the Red Crag area this bed has suffered so generally and extensively from the wash of the sea during the emergence of the country, when the Middle Glacial gravel (No. 8) was in course of accumulation, and the land ice, of which the chalky clay was the moraine, was extending from the Wold to follow the retiring sea, that only patches of it remain there. One of these patches, that at Kesgrave, is shown in the map, but another occurs at California-by-Ipswich, another at Kirton, and another at Rookery Farm, Eyke, none of which are shown in it. All of these appear to be of considerable thickness (40 to 50 feet), and the first and last of them have a little of the Middle Glacial gravel over them in places. Another patch, on the Hasketon side of Woodbridge, is overlain by the chalky clay; and at Tuddenham, near Ipswich, the base of this brickearth is exposed passing down into the sand No. 6, of which about twenty feet underlies it, and rests on the London Clay; and there also the denudation of this brickearth, which took place prior to the deposit on it of the Middle Glacial gravel, is well shown by the irregular way in which that gravel lies upon it. Remnants occur also in other parts of South Suffolk, but they are beyond the limit of the map.[1] In the Section (A) drawn through the Red Crag area, the Middle Glacial is therefore erroneously represented as resting generally on the Red Crag, whereas this is exceptional, and the Lower Glacial sands should have been shown in most parts (i.e. in those where they have not taken the place of the Crag altogether) as intervening, and the thickness of the Middle Glacial been there proportionately reduced. The correct position of all the beds of this sequence is shown in fig. 1 of the plate which accompanies my memoir on the "Newer Pliocene Period in England," in the thirty-sixth volume of the 'Quarterly Journal of the Geological Society,' the line of which is drawn through three of these remnants of the brickearth; and in it the Middle Glacial gravel is shown on the plateaux as very thin, and in places absent altogether, but as thickening towards the brows of the valleys, which, when they were in the condition of troughs excavated in the rising sea bottom of the sand No. 6, had been filled by it; the gravel in the central parts of these troughs having been cut out as these were deepened by the shrinkage into them of the ice of the chalky clay, or by the action of the sea, as emergence went on. A well which I sunk to a depth of eighty-four feet subsequently to the publication of that figure, but on the exact line of it, and on the eastern edge of the plateau from which the valley of the Deben is cut down, showed this gravel to be there seventy feet thick beneath six feet of the chalky clay (the upper thirty feet being full of the chalk *débris* of that clay), and that the sands No. 6 had been almost all removed to give place for it. It is this sand, or else that formed by the decalcification of the Crag, and not the Middle Glacial, which overlies the Crag shown in the cut on page xxi of the "Introduction" and in Sections XIX and XX.

The map thus requires to be corrected by the intercalation of a belt of the shade and colour representing the sand No. 6 between the Red Crag and the Middle Glacial; and it

---

[1] One of these, at Stowmarket, is in the footnote to p. 22 of the "Introduction," referred to as of post-glacial age, and another about six miles north of Ipswich, and three-quarters of a mile south-west of Hemingstone Church, is shown in the map by a dot of the wrong colour (that of bed No. 10). I am informed also by Mr. Dalton, of the Geological Survey, that he found an exposure of this brickearth under the chalky clay at Baddingham, just midway between the patch of it shown in the map at Bloxhall, in South-east Suffolk, and the exposure of it at Withersdale, on the Waveney, near Harleston, so that probably much of the chalky clay of High Suffolk is underlain by remnants of the same bed.

also requires the substitution of this colour for that of the Middle Glacial over most of the area east of the chalky clay, which stretches from Sizewell to the River Blyth, and to the cliffs of Easton Bavent and Covehithe; there being but very little, if any, of the Middle Glacial present over this area, which is occupied by the sand and shingle No. 6 in greater thickness than elsewhere.

The Section (R) of Dunwich Cliff, and that (s) of Easton Bavent and Covehithe Cliffs, also require correction, the bed shown in the latter as the Contorted Drift (No 7) being the same as the capping loam of Dunwich Cliff, which in Section R is shown under the number 10;[1] both of them being, as a late examination of them has enabled me to perceive, a morainic bed formed (in Dunwich and the southern part of Easton Cliffs, from a reconstruction of the pebbly sand No. 6 with some admixture of the material of the chalky clay, and in the northern part of Easton Cliff, from a reconstruction of these sands and the Chillesford clay together,) by the ice in its passage to the sea after this part of Suffolk had emerged towards the close of the chalky clay formation; and the gravel, shown by the number 10, as resting on this bed and on the Chillesford clay in this cliff, and shown also in Covehithe Cliff, is merely a part of this morainic bed, being pots of pebbles derived from No. 6. A bed of this morainic material cutting like a dyke through the sands No. 6 at the southern end of Easton Cliff (where this cliff is only six or seven feet high) requires to be added to Section s. Another such bed forms the northern extremity of Southwold Cliff, overlying the bed of derivative shells in the shingly sand No. 6, presently to be referred to. The section of Dunwich Cliff also requires correction by the omission of the Middle Glacial which is shown in it under the numbers 8″, 8‴, and 8⁗; all of this being part of the sand No. 6, to which the shingle under the ruins (shown in Section R by the figure 10) also belongs; and this shingle is still more largely present in that sand at the southern end of this cliff. The whole of Dunwich Cliff, from below the beach line up to the capping loam of morainic origin just mentioned, is thus formed of No. 6, the intercalation of clay shown in Section R by the figure 9 being probably a modification of the sandy formation, by the introduction of argillaceous material analogous to that which gave rise to the Cromer Till and Contorted Drift of North Norfolk; both of which are, in my view, merely modifications of the same shingly sand by the introduction of a different sediment.

Descending thus below the beach line, and forming (with the morainic loam already mentioned) the whole of the cliffs of Dunwich and Southwold, this sand there occupies a space from which the Chillesford clay and the upper part of the Crag beneath it had been removed, so as to form a channel in the Lower Glacial sea which divided two islands formed of Chillesford clay and Crag beds; of which islands the southern was comprised by the country extending from Butley and Chillesford to Sizewell, and the northern by the area of which the cliffs of Easton and Covehithe (Sect. s) furnish a section. The sands No. 6, which, as already mentioned, cover the Red Crag area, lie up to the

---

[1] See the footnote No. 5 to p. 29 of the "Introduction."

southern of these two islands, as well as extend over it, just as they do in the case of the northern, and so that, being bedded in the channel and up to the shore of this southern island, they lie much below the level of the Chillesford beds which cap it at Chillesford, Sudbourne, Iken, Oxford, and Aldboro', as well as below much of the Coralline and Red Crag on which those beds there rest, and of which that island is formed.[1] Occupying also the channel dividing these islands from each other, and in that way furnishing the section of Dunwich and Southwold Cliffs, these sands lie up to the shore of the northern island thus formed of beds of Crag age, as may be seen in the southern part of Easton Cliff when this is sufficiently free from talus. It is in this part that a bed of shells occurs in these sands, and it is the only one, so far as I am aware, that they yield in Suffolk. This shell bed is exposed at the northern end of Southwold Cliff, about the beach level, and immediately under the morainic loam already mentioned;[2] and I call attention to it because I believe that *all* the shells in it are derivatives from the Crag of which this Lower Glacial island was formed, before the progress of the submergence overwhelmed it, in a similar way to that in which so large a part of the shells in the Red Crag are derivatives from the island and peninsula of Coralline Crag which existed in the Red Crag sea. Not only is the characteristic species of these sands in Norfolk, *Tellina Balthica*, not present in this bed, but the shells that are in it, even the strongest, such as the Littorinæ, are for the most part fragmentary. The shells which I was able to detect in it during many repeated searches were the following, viz. *Nassa incrassata, Purpura lapillus, Cerithium tricinctum, Turritella*

[1] The southern of the two islands mentioned in the text may have been divided into three smaller, by channels now represented by the mouth of the Alde and by the Butley creek, in which these Lower Glacial sands may have been bedded and since removed; for at Iken Cliff, on the Alde, these sands are in section at the sea level, nearly fifty feet below the contiguous top of the Chillesford beds on this island. This southern island (or islands) was probably abutted on the south by another island formed of Red Crag, and now buried beneath the Lower Glacial sand (capped with more or less of the Middle Glacial gravel) of the heaths of Hollesley, Boyton, Sutton, and Alderton; for exposures of Red Crag along the edges of the small valleys penetrating this tract of country occur at as high or even higher level than the Chillesford beds just referred to. This, again, was probably divided by a channel now represented by the Deben from another island of Red Crag, represented by the tract between the Deben and Orwell estuaries, and this again by one represented by the tract between the Orwell and Stour estuaries; as from the way in which the Lower Glacial sands take the place of the Crag in many parts along the sides of the valleys of these estuaries, these latter may very likely have been channels during the earlier part of the Lower Glacial sea, and been once filled by its sands, which were removed by the action of the sea, followed up by the land ice as the land was emerging during the formation of the chalky clay. Whether the Chillesford clay ever was spread out over that part of the Red Crag which occupies the area between Butley and the Stour, and was afterwards removed, or whether this southern part of the Red Crag was land during the slight depression under which the Chillesford beds were spread out, there are no means of determining, though the Chillesford clay seems to have been deposited in north-east Essex (Walton), and up the Gipping valley at Needham.

[2] This bed was also found about half a mile inland in making the railway cutting near Southward station.

terebra, Littorina littorea, Natica clausa, Leda oblongoides, Lucina borealis, Cardium edule, Astarte compressa, Cyprina islandica, Tellina obliqua, Corbula striata, and *Mya arenaria;* all being species which occur in the adjacent Crag beds.

The fluvio-marine Crag from which the Chillesford beds have been removed to form this channel, and on which the sands No. 6 thus rest below the beach line, comes through the beach in two very small knobs about a quarter of a mile from the southern end of Dunwich Cliff, which are crowded with shells; and it yielded me also an equine tooth.

Lastly, I have in the memoir of the "Newer Pliocene Period" in England, already referred to, given my reasons for regarding the Bridlington bed from which the Mollusca given in the "Upper Glacial" column of the tabular list at the end of the first Supplement to the "Crag Mollusca" were obtained, and also the basement clay of Holderness with which that bed is associated, as being of Lower Glacial age, such clay being, in fact, the actual moraine of the ice from which proceeded the material interstratified in the Cromer Till (No. 6 *a* of the Map, &c.); and for regarding the molluscan remains given in the "Middle Glacial" column of the same tabular list, as being an admixture of remains from the bottom of some fiord which had been in process of accumulation from the commencement of the sands No. 6, and during the whole of the Glacial submergence, but which was ploughed out by the ice of the chalky clay during its advance as it followed the retreating sea during emergence; so that these remains became embedded by this derivative process in the upper part of the Middle Glacial (No. 8 of the Map and Sections), as that bed was emerging, and just before the chalky clay moraine was pushed over it.

I should add that though, to avoid confusion in this explanation, I have adhered to the term Middle Glacial, this formation is (in the view to which the continued study of the subject has brought me) merely the marine accumulation which was synchronous with the moraine of the land ice which is represented by the chalky clay; and the precise mode in which the two were accumulated, according to my view, is traced in detail in the memoir just referred to.]

PLATE I.

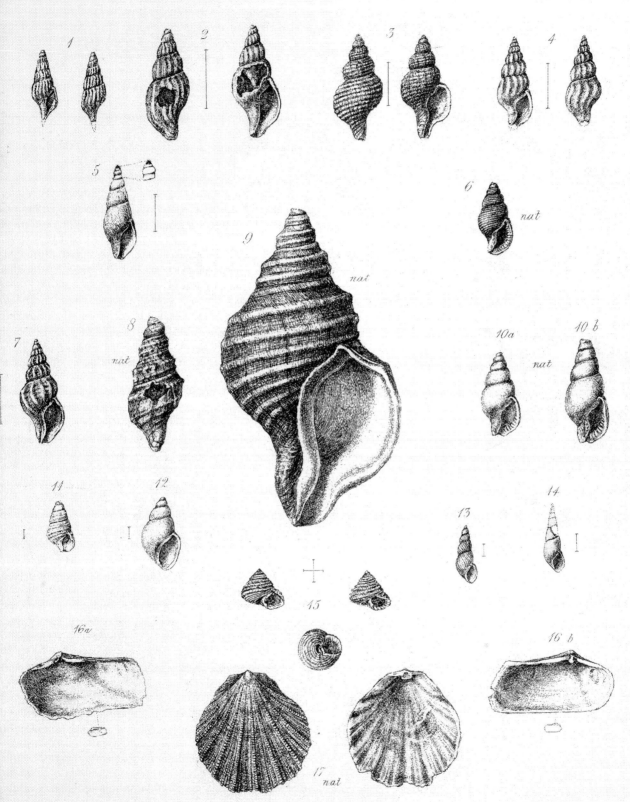